Vinyl Records and Analog Culture in the Digital Age

Vinyl Records and Analog Culture in the Digital Age

Pressing Matters

Paul E. Winters

LEXINGTON BOOKS
Lanham • Boulder • New York • London

Published by Lexington Books
An imprint of The Rowman & Littlefield Publishing Group, Inc.
4501 Forbes Boulevard, Suite 200, Lanham, Maryland 20706
www.rowman.com

Unit A, Whitacre Mews, 26-34 Stannary Street, London SE11 4AB

Copyright © 2016 by Lexington Books

All rights reserved. No part of this book may be reproduced in any form or by any electronic or mechanical means, including information storage and retrieval systems, without written permission from the publisher, except by a reviewer who may quote passages in a review.

British Library Cataloguing in Publication Information Available

Library of Congress Cataloging-in-Publication Data
Names: Winters, Paul E.
Title: Vinyl records and analog culture in the digital age : pressing matters / Paul E. Winters.
Description: Lanham : Lexington Books, [2016] | Includes bibliographical references.
Identifiers: LCCN 2016021956| ISBN 9781498510073 (cloth : alk. paper) | ISBN 9781498510080 (electronic)
Subjects: LCSH: Sound recordings—History. | Sound recordings—Social aspects.
Classification: LCC ML1055 .W57 2016 | DDC 384—dc23 LC record available at https://lccn.loc.gov/2016021956

ISBN 9781498510097 (pbk : alk. paper)

☉ ™ The paper used in this publication meets the minimum requirements of American National Standard for Information Sciences Permanence of Paper for Printed Library Materials, ANSI/NISO Z39.48-1992.

Printed in the United States of America

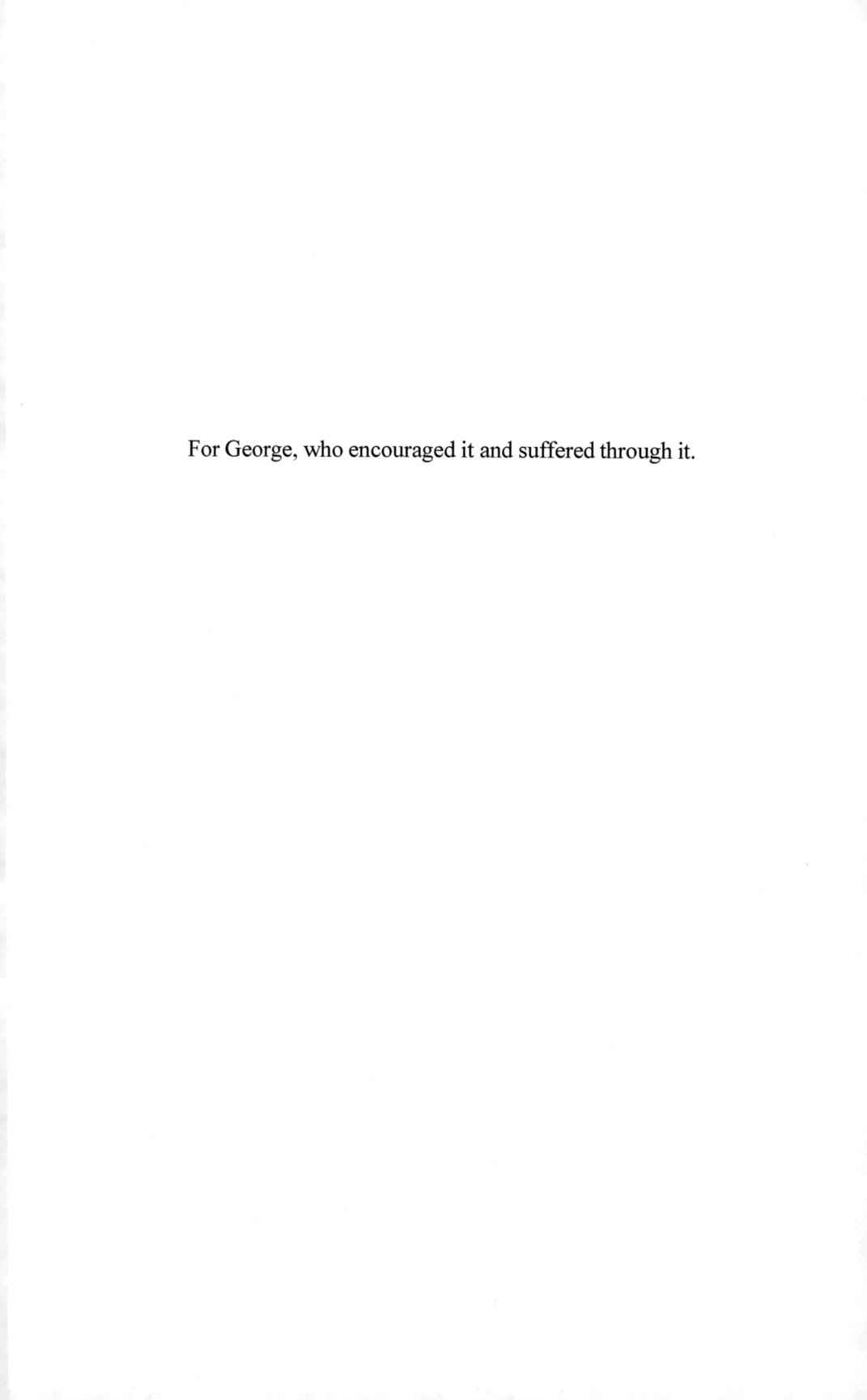

For George, who encouraged it and suffered through it.

Table of Contents

Acknowledgments ix

Cueing Up xi

1 "Dogs Don't Listen to Phonographs": Nipper, "His Master's Voice," and the Discourse of "Fidelity" 1

2 The Beatles on itunes and Vinyl Reissue: Aesthetic Discourse and the Listening Subject 23

3 Virtual Authenticity: The Reemergence of Vinyl in the Digital Age 45

4 Criminal Records: Record Collecting as Counter-Discourse 61

5 "Cabinets of Wonder" or "Coffins of Disuse?": Reissues, Box Sets, and Commodity Fetishism 87

6 "You Spin Me Round (Like a Record)": Analog Audiophilia as Disciplinary Mechanism 109

7 "The Vinyl Anachronist": The Role of Social Media in the Formation of Communities of Vinyl 133

The Run-Out Groove 159

Bibliography 161

Index 175

About the Author 191

Acknowledgments

This book would never have happened were it not for the Popular Culture Association (PCA) and Lindsey Porambo of Lexington Books. Chapters 1 through 3 started life in much altered form as papers I delivered at the conference from 2012 to 2015. Thank you to Thomas Kitts for accepting them and for the kind indulgences and the intelligent reactions of the conference participants to my work. Thank you also to the music group of the PCA for the many intellectually stimulating papers and discussions they have participated in. Thanks also to Dr. Kitts for introducing me to Mark Volman in New Orleans in April of 2015! Thank you to Lindsey Porambo for suggesting that this could be the making of a book and for her forbearance she showed me through the process of getting it to draft form.

I thank my colleagues at DeVry University, North Brunswick, NJ, for their support and encouragement. I would like to thank two especially. Anca Rosu, who shares my interest in the relationships between people and technology if not in popular music, has been a good, supportive friend and an excellent sounding board for my ideas, even providing moral support for my conference presentations, although they were not in her area. For all of this I thank her. Also, our librarian, Joe Louderback, has helped me chase down more than one source in the course of writing this book. I thank him for all of his help. My students have also provided me with constant intellectual stimulation and helped me to formulate and to refine my ideas. When I started teaching at DeVry in 2000, most of my students would not have known what vinyl records were, or at least they would have thought of them as something their parents or grandparents used. Now when I bring the topic of vinyl records and turntables up in classes, most of the students know what I'm talking about, a welcome change.

Thank you to my teachers, mentors, and guides at Lehigh University, particularly Rosemary Mundhenk, Alexander Doty, and David Hawkes, for helping to provide me with the theoretical foundation to come to terms with culture and the ways human beings and technology interact and influence one another. In particular, I would like to thank Carol Wells, who I met on my first day of graduate school and who has been my friend, offering intellectual and emotional support, ever since. She also told me that writing a book on vinyl records would make me "the king of the hipsters," but I think I will have to wait for the coronation. I would also like to thank the members of the YouTube Vinyl Community, who gave me their time and shared themselves and their interests with me. Seeing the ways in which vinyl was emerging online provided me with the original impetus for thinking and writing about this topic to begin with. Finally, thank you to my family who contributed to all that is good in me. Thank you, John, Mark, Dianne, David, and Blaine and all of your spouses, children, and grandchildren. I hope that my late mother would have been proud of her son and have thought that all of the trouble she thought I was putting into my record collection was worth it. Finally, thank you to my husband George and his family for their love, support, and acceptance. Anything in this book worth reading and retaining, all of these people have contributed to. I claim sole credit, however, for any errors and inaccuracies.

Cueing Up

Undertaking a study of the reemergence on vinyl records and of analog technology in the digital age has proved to be challenging in many ways, not the least deciding how much history the subject should include, figuring out how to talk about vinyl technology separately from other sound technologies and how and where those other technologies needed to be addressed, and adequately contextualizing this discussion in the field of sound and technology studies in general. Even more of a challenge, however, was the constantly shifting substance of the object of study brought about by new developments. It sometimes seems as if every day there is some new technological, social, or cultural development being reported about vinyl records. Stories would literally appear online and in the media about vinyl records as they were being written about, which of course would profoundly affect what was being said or change the argument significantly. For anyone whose interests lay with the remarkable fact of vinyl's second act on the technological stage, though, the explosion of attention that culture is currently paying to vinyl's resurrection can only be seen as a positive thing.

And, if vinyl records are a niche technology in 2015, they are a niche that has grown substantially from the low point of the mid-2000s, and the niche continues to grow, even if it does not match the falling sales of CDs and digital downloads. Glenn Peoples and Russ Crupnick make this point in an article called "The True Story of How Vinyl Spun Its Way Back From Near-Extinction," in which they discuss music industry's attempts to find a replacement technology for the fading CD in 2006–2008 (Peoples and Crupnick 2014). Peoples and Crupnick say,

> Vinyl concepts were tested in 2006 next to early versions of connected CDs and CD/DVD hybrids. This version of vinyl included a digital download card that would allow the buyer to download the album's digital tracks. Of all the

> products tested between 2006 and 2008, vinyl tested the worst. There were ample problems. Vinyl had the lowest purchase interest. It didn't particularly appeal to any important fan segments. And it wasn't seen as a valuable item. More people expected to pay less for vinyl than were willing to pay a premium. (Peoples and Crupnick 2014)

Despite what their research told them, however, listeners in the intervening years have proven it wrong. Peoples and Crupnick note that "Since the breakout year in 2008, vinyl albums sales increased 223 percent to 6.06 million units last year. And rather than slow down, sales are actually gaining momentum. In unit terms, vinyl sales grew 1.5 million units last year. With sales on track to surpass 8 million units, this year's growth will surpass 2 million units" (Peoples and Crupnick 2014). They attribute the change to Record Store Day (RSD) being established in 2008, and to some degree they are correct, but the online vinyl revival was already a going concern by the time of the first RSD, and while it certainly did much to raise the profile of vinyl records in the public consciousness, the independent retailers and record labels that helped make RSD happen were actually catching a wave that was already moving.

An article in *The Oregonian* notes that "[v]inyl's popularity faded in the '80s, as the CD displaced the format with the digital promise of 'perfect sound forever,' but it wasn't entirely abandoned. Per SoundScan, the sales-tracking system that began in 1991, sales of new vinyl LPs were actually on the rise in the '90s, heading from 625,000 in 1994 to a peak of 1,533,000 in 2000" (Greenwald 2014). All the way back in 1998, Steve Threndyle could write,

> For most people who listen to music, news that vinyl records are making a comeback might invite disbelief—"Sure, and I suppose we'll all be typing on manual typewriters and watching black-and-white TV, too." But it's true: turntables, vinyl records and even tube amplifiers are staples of what might be called an analog hi-fi renaissance. (Threndyle 1998)

Perhaps Threndyle's pronouncements of a renaissance were a bit premature in retrospect, but they do go to show that vinyl's comeback was being discussed in the press long before the first RSD occurred. While success may have been later in coming, however, when it did arrive it was in a way much larger than Threndyle might have predicted in 1998. Toward the end of 2015, it would be reported that vinyl sales had surpassed the revenues generated by digital streaming, even if those numbers neglected to account for the money generated by digital subscription services (Wolff-Mann 2015). In 2014, Neil Shah could coin the aphorism, "[t]he record-making business is stirring to life—but it's still on its last legs" (Shah 2014). Shah's point is that the vinyl record industry is to some degree a victim of its own success, a success that was not foreseen and therefore not planned for. Aging pressing plants and machines and shortages of vinyl have led to "bottlenecks" in the supply line, particularly when they contend with Record Store Day and "large orders from superstars" (Shah 2014). The result in real terms is the reopening of shuttered factories. Shah also notes of the machines that "record-plant owners have been scouring the globe for mothballed presses, snapping them up for $15,000 to $30,000, and plunking down even more to refurbish them" (Shah 2014).

Also remarkable is the fact that well into the second decade of the twenty-first century new record stores continue to open all around the world. Bryn Lovitt reports that "[a]t the official 2015 Record Store Day launch event at Brooklyn's Rough Trade Records today, RSD co-founders Michael Kurtz, Eric Levin and Carrie Colliton were happy to report that indie record shops are continuing to make a comeback. 'Two-hundred independent record shops are opening this year,' said Kurtz" (Lovitt 2015). And, while 200 new shops might not seem that large a number, Lovitt makes the point that it is "still a robust number considering that just a few years ago, the rise of retail behemoths in the music industry and then the ubiquity of digital distribution (legal and otherwise) threatened the small indie shop's very existence" (Lovitt 2015). Also, Greenwald makes the point that "the official numbers don't tell the full story." In addition to leaving out many indie shops that do not participate in SoundScan, the "big revival might be a lot bigger because SoundScan doesn't track sales of used vinyl, long the domain of crate-diggers, vinyl newcomers and flea-market collectors alike" (Greenwald 2014).

Influenced by postmodern theory and cultural studies, this book attempts to call the technologies surrounding vinyl records to account for the ways they influence and are influenced by social formation and popular discursive practices. Of many influences on its understanding of the way that culture and technology operate and influence each other, this book must particularly acknowledge a few studies that preceded it and provided it with grounding influence. Among many other important studies, the work of Lisa Gitelman and of Jonathan Sterne is especially important to any endeavor that attempts to come to terms with sound technology in its historical context. Gitelman's *Scripts, Grooves, and Writing Machines* and Sterne's *The Audible Past* in particular proved highly influential. William Kenney and Evan Eisenberg have both written important works about the history of recorded music and its cultural significance, and their work should thus be acknowledged as foundational. Finally, as this book was being written, *Vinyl: The Analog Record in the Digital Age* by Dominick Bartmanski and Ian Woodward was published, a similar study to this one that has proven helpful in many ways. Also of note are David Hesmondhalgh's sociological study of music *Why Music Matters* and Travis Elborough's examination of the album *The Vinyl Countdown*. Singling out these significant works, however, should in no way be taken as a slight to all of the other important works that have contributed to this book's understanding of analog sound technologies at this particular historical moment. All of these works attempt to come to terms in one way or another with what makes vinyl records and analog technologies important to human beings and how they come to terms with them. What follows should be read in a similar light.

Chapter 1 begins this study at roughly the beginning of the era of analog recording, examining notions of fidelity, how they emerged in the late nineteenth and early twentieth centuries centered on discursive practices surrounding the development and the domestication of the record player, particularly as such notions are embodied in "Nipper," the advertising trademark of the Victor, and later the RCA Victor corporation. Chapter 1 argues that the fidelity of the recording as a reflection of the human voice was and is an important perception to develop for the public's understanding and acceptance of recorded music as a "thing," as a medium. Using Walter Benjamin's notion of "aura" from "The Work of Art in the Age of Mechanical Reproduction" as a jumping point, chapter 2 examines recorded music as an aesthetic discourse, using as its exemplar the music of perhaps one of the most important recording acts of the twentieth century, the Beatles. It considers the ways that notions of what makes a sound recording superior and what makes for the best medium by which that sound recording is disseminated are entangled even as they provide warrant for each other's claims and the ways in which a recording's imperfections play a part in its aesthetic valuation.

Chapter 3 looks at the ways in which the analog record revival is bound up in the discourse of "authenticity," even if there mostly appears to be little ontological grounding for such claims. Chapter 3 rather sees authenticity to be more a productive trope, along the lines of Raymond Williams's "structure of feeling," through which listening subjects build and understand their realities in the face of the perceived disconnections and incorporeality of the twenty-first-century digital realm. Chapters 4 and 5 both center on collecting practices and the ways in which they influence and are influenced by the perceptions and understandings of vinyl record collectors. Chapter 4 looks at record collecting as a counter-discursive practice to the mainstream discourses of the recording industry. Record collecting not only serves to gather and to preserve recorded music history, but it also does so in the face of powerful discursive practices that would orient listening subjects more to what is new than to what has been in the past. Chapter 5 notes the ways in which the recording industry co-opts and reinscribes the discourse of collectors in order to reestablish them as consumers and their objects as fetishized commodities. Collectors may have been responsible for keeping vinyl alive and bringing it back from the dead, but it is the work of the recording industry, with its offerings of elaborate packages and audiophile reissues, that could be seen to be responsible to some degree for its continued vitality and growth.

Chapter 6 considers the ways in which the playing of records, in particular in the register of what might be referred to as audiophilia, provides a productive discipline that assists in the construction of the listening subject, including the (gendered?) ways in which he or she conceives of himself or herself. To some degree, the mastery of the complexities of the analog recording and playback mechanisms contribute to self images of individual listeners, making the difficulty of those technologies to get it "right" a not insignificant part of their appeal. Chapter 7 examines in detail one of the foundational claims of the entire study, the role of information technology, more specifically the Internet and social media, in the reemergence of vinyl records and analog playback technologies as a viable medium for music in the twenty-first century. Online communities of vinyl listeners have sprung up all over cyberspace, as individual listening subjects gather together to share and to compare their passions, using all of the tools that online publication offers. The self-fashioning that social media allows also helps to construct vinyl or analog listeners as a distinct identity in the virtual realm, one that can be said to spill over into the "real" world as we perceive it. This book argues that if the consumption and collecting of vinyl records has grown beyond a cult in the last two decades, it is, sometimes paradoxically, both an outgrowth of and a reaction to the explosion of digital media and the information revolution.

This book will make no aesthetic claims about the superiority or the inferiority of one technology for the delivery of recorded music or another, be they analog or digital. Rather, this is a book *about* aesthetic claims that are made about these technologies, and it seeks to explore in one way or another how such claims come to be made and on what epistemological claims they might base themselves. It presents its argument regarding these claims not in a linear fashion, but rather assembling its claims from what might be described as various angles of approach to the same question, that is, why, despite the best efforts of the recording industry and its confederates in the electronics industry, the consumption and collection of vinyl records not only never went away, it continues to grow and to thrive in the new millennium. This book does not seek, nor should it read as presenting, any claim that might be inferred as either comprehensive or universal in answer to the questions it raises about analog sound technologies and the subcultures they engender. Instead, it offers an archeology in the Foucauldian sense of the ways in which the "listening subject" constructs his or her identity in relation to both analog and digital technologies as categories of social relations.

Nor should this book be taken for a sociological study of record collecting and record collectors, although it does partake of some of the insights of sociological explorations of technology and social formation such as the work of Sherry Turkle, and it also includes some limited field work on social media and the self-identification of vinyl enthusiasts. Which is not to say that such work would not provide valuable insights. If this study does its work, however, it will provide a theoretical foundation that might ground further study of these phenomena, psychological, sociological, philosophical, historical, and aesthetic. This book should be considered an analytical examination of some of the various engines that drive the resurgence of vinyl as a viable medium for the dissemination of music and sound. As such, it partakes of some of the insights of all of the aforementioned disciplines, but should mostly be seen as an attempt to apply literary, post-modern, and cultural theory to to its objects of study. If this book is successful, it will provide some insights, based on analytical tools provided by the study of literature and culture, into why vinyl has come back practically from the grave, and why it has particularly at this perhaps incongruous historical moment. To the degree this study gets it right, it should give credit to the aforementioned works; if this book gets it wrong, that should be seen as entirely the fault of the author.

Chapter One

"Dogs Don't Listen to Phonographs"

Nipper, "His Master's Voice," and the Discourse of "Fidelity"

In *Truth and Method*, Hans-Georg Gadamer says that the "question of the truth of art forces us, too, to undertake a critique of both aesthetic and historical consciousness, inasmuch as we are inquiring into the *truth* that manifests itself in art and history" (Gadamer 1988). As many of us return to the vinyl fold, it may be worth our whiles taking a moment in this context to examine that to which we are returning. For this examination, it might behoove us to turn our attention back to the days of shellac, to 1899, in fact, and a familiar image, titled "His Master's Voice." Before the image of the dog listening to the gramophone transformed into an instantly recognizable advertising icon, it was a painting. By all accounts, the painting depicts an actual dog, called "Nipper," the pet either of the painter, Francis Berraud (Goodrum and Dalrymple 1990), or of his late brother, Mark (Designboom 2000–2010). Berraud initially couldn't sell the painting. In fact, he was rejected for exhibit by the Royal Academy; being told, "no one would know what the dog was doing" (EMI Records Ltd. 1997). The contention was that viewers would have little frame of reference even for humans listening to phonographs, let alone dogs! Early twentieth-century viewers of this image needed to be taught how to read it, what Nipper was doing, and in the process of learning to read it learn how to do it themselves, a process to which we will return later.

The painting "His Master's Voice," however opaque it might have been to its contemporary viewers, does exist in a tradition of Victorian animal painting and narrative of which the contemporary viewer might have made sense. In "Dog Years, Human Fears," Theresa Magnum notes that in the late

Victorian era, there began a preoccupation with animal, and particularly dog, subjectivity that she relates specifically to concerns about old age and mourning in human beings (Magnum 2002, 35). She says, "The history of the dog in nineteenth-century Britain is itself a tale of identity formation" (Magnum 2002, 36). Magnum further contends, "[a]nimal subjectivity is articulated with special force in visual representations of dogs" (Magnum 2002, 37). This subjectivity finds its own particular expression in the act of mourning, which, according to Magnum, posits the animal as an "increasingly loaded signifier for its own previously unarticulated self—a conscious canine capable of the intricate, textured memories, the deep melancholy, and the variegated imagination that inspires lonely visions of a solitary future" (Magnum 2002, 38). Magnum turns specifically to Berraud's painting in making this point, and reading Nipper's image as a representation of the subjectivity of grieving, she states,

> Berraud claimed that when he played his dead brother's voice to his brother's pet terrier, Nipper, the dog would strike a puzzled pose of longing before the gramophone. Ears half-cocked, face drawn down by sagging, wrinkled jowls, Nipper attempts to smell as well as to hear his lost master. What twentieth-century viewers interpreted as a dog puzzled by the marvels of technology, Victorians perceived as a poignantly longing canine subject, eager, perhaps, for the voice of his master and for a voice with which to express inarticulate, unending faithfulness and grief. (Magnum 2002, 38)

If Magnum's interpretation of how a late-Victorian viewer would have read Berraud's image holds true, then it makes for quite a morbid mascot to represent the Victor Talking Machine Company. Perhaps, then, the canine subjectivity that the image represents is itself a metaphor for another type of subjectivity, one that exists at the intersection of recording technology, identity formation, and discursive practice, that of what I will call "the listening subject."

In order to unpack that metaphor, it might be useful at this point to examine the components of "His Master's Voice" in order to see what sense a critical analysis might make. In addition to being a sign of "fidelity," Nipper could also be perceived to be a symbol of "domestication." Domestication theory holds that new and unfamiliar technologies need to be "tamed" upon their introduction into the home, and that this training process transforms the technology, the home, and the actors. Roger Silverstone makes a point that might be useful to our understanding of Nipper's role in how the phonograph became a fixture of the household furnishings. Silverstone defines domestication as "a process of bringing things home—machines and ideas, values and information—which always involves the crossing of boundaries: above all those between public and private, and between proximity and distance, is a process which also involves their constant renegotiation" (2006, 233).

In fact, in their collection of essays on domestication theory, *Domestication of Media and Technology*, Berker, et al. even use a canine metaphor, comparing technologies at one point to a "young puppies," to make their point about the domestication of technology, saying "'strange' and 'wild' technologies need to be housetrained; they have to be integrated into the structures, daily routines, and values of user and their environments" (Berker et al. 2006, 3). Nipper's introduction as an advertising mascot was part of an ongoing attempt on the part of the nascent recording industry to move the talking machine, phonograph, and gramophone out of the public sphere and into the private one. The introduction of these machines into the public sphere in the late years of the nineteenth century came via penny arcades and nickelodeons, spaces that carried with them a sense of the tawdry and the disreputable, also reflected in the recordings that listeners could hear by dropping their nickels into the machines (Kenney 1999, 43).

Perhaps more than any individual, Eldridge Johnson was responsible for the domestication of the talking machine into the American home. Johnson, the director of the Victor Talking Machine Company, along with other early innovators such as Thomas Edison and Emile Berliner, established a policy of reflecting values of home and hearth as a means of taming recording technology for entry into the late Victorian home. However, added to the bad reputation of the early record player, taming the technology also meant making it more user-friendly for effective home use. Alexander Magoun says about this problem that it "required negotiations between inventors, entrepreneurs, and consumers over how to make a commercial product—a relatively fragile wax cylinder and its relatively complex playback equipment—into a part of the average household, subject to the stresses of usage by far more people less trained in operating the system" (Magoun 2000).

"Improvement" of the listeners of this new machine was also on the agenda. William Kenney states that such improvements

> were guided by the Victorian era's association of the home as an "oasis of calm" at which the wife/mother provided, among other things, refined and uplifting music with which to rejuvenate her hard-working husband and edify, enrapture, and improve the memories of her children, imparting a sense of proportion, good taste, high moral purpose, and brotherly and sisterly affection through inspiring music. (Kenney 1999, 46)

Along with Victor's introduction of the Red Seal label, with its connotations of taste and refinement, came innovations in advertising, Nipper among them. Of Johnson's advertising acumen, David Suisman says, "If the opera and Red Seal advertisements imbued Victor phonographs with drama, seriousness, and respectability, the trade-mark dog evoked complementary ideas of loyalty, amusement, and domestication . . . He was trained, groomed, and well-behaved. He was allowed in parlors" (2009, 118). Bringing the record player along with the dog into the parlors of the middle class was very much on Johnson's agenda.

Of course, there is the ideal, and there is the practice of individual households and users. John F. Sherry, Jr. notes of iconography in advertising that the "polysemous nature of the symbol makes it ideally suited to the manipulating of ambiguity which is the hallmark of much advertising" (Sherry 1987). The dog is never really fully trained and retains some of the wolf in its very DNA. "Taming" technology is only ever partial, and it requires constant renegotiation in order to retain its place in the parlor (Berker et al. 2006). Nipper is much more in this sense than an image of mute uncomprehending loyalty (Suisman 2009, 118). He is also more than a symbol of the power the phonograph had over its listeners. William Kenney says, "In every case, the phonograph's power, wielded by white, middle-class and upper-middle class owners, transfixed the weak and disenfranchised" (Kenney 1999, 55). We should resist the temptation to see Nipper as this sort of symbol of passivity. While only a few entrepreneurs may have held the patents on the technology and therefore maintained tight control of the market (Kenney 1999, 55), the industry eventually grew far beyond any of their expectations and in ways none of them could have begun to imagine. Furthermore, while some of the corporations they established to help keep their grip tight remained, for the most part, the industry imagined by Edison, Berliner, Edward Easton, and Eldridge Johnson had transformed beyond their recognition by the 1920s. The Victorian parlor they had hoped to bolster became the twentieth-century recording industry, transforming culture as it was itself transformed (Kenney 1999, 64).

We should remember in this context that "His Master's Voice" was a work of "painting" before it became a work of "publicity," to use Thomas Berger's terms (1973). As a visual image, Nipper is also an object of the gaze, which places him in the complex matrix of power relations involving commodification and consumption, or, as has been said of Foucault's work on the gaze, "The gaze is integral to systems of power and ideas about knowledge" (Sturken and Cartwright 2009, 94). As John Berger notes, having "a thing painted and put on a canvas is not unlike buying it and putting it in your house. If you buy a painting you buy also the look of the thing it represents. This analogy between possessing and the way of seeing which is incorporated in oil painting is a factor usually ignored by art experts and historians" (Berger 1973, 83). Is "His Master's Voice" a painting representing the power the technology has on the trained animal, or is it perhaps more complicated than the image suggests? The image becomes a symbol of the consumption of the thing it represents, which in this case is the act of listening.

The gaze, furthermore, raises yet another problematic question, the question of identification. With whom or what is the viewer/listener to identify in this image? The dog is mute and misunderstands that which he sits before. Is his misunderstanding *the viewer's* misunderstanding? Is his mute listening posture a metonym for the entire recording apparatus? It is difficult to determine that with which the viewer *should be* identifying. The title of the image might have offered clarification, but it only serves to make matters worse. Am I the dog, incapable of comprehending the voice I hear and "recognize?" Am I the master, self-assured in the recognition of my own voice, even by a dumb animal? Am I the voice itself, transcending space and time in a technological paradise? To compound the question, the Victor Talking Machine that this image purports to represent offers only playback capability, so the voice of the master would need to be a contracted performer, not the home listener.

If the "puppy" still needed training in order to become housebroken, here perhaps is the paper with which to start. Around 1916, customers buying cylinders were given a sheet of paper containing a complicated set of instructions for what was called the "Edison Realism Test" (Katz 2010, 23). This "test" required listeners to try to empty their minds of extraneous information in preparation for a comparison between the recorded performance they were currently listening to and the last time they witnessed a similar performance (it isn't specified, but there is a strong suggestion that recorded and live music are being compared in this test). While Mark Katz cites this test in order to make a point about the visual aspects of sound performances (2010, 22), a more interesting point in light of the current discussion is the realism test's following description of the desired result: You should get the same emotional re-action experienced when you last heard the same kind of voice or instrument.

> If you do not obtain this re-action at the first test, it is due to the fact that you have not wholly shaken off the influence of your surroundings. In that case you should repeat the test until you are no longer influenced by your surroundings. (The Edison Realism Test n.d.)

In other words, if the listener does not achieve the desired result, the problem isn't with the test or with the technology, but rather it is *the listener* who has the problem! The "Edison Realism Test" seems more like training, panoptic in a Foucauldian sense, training the listener what to listen for and recommending the policing of the self for evidence of the test's failure. "Fidelity" not only trains listening subjects in what to hear, it also trains them in what they should be listening *for*, what to expect and anticipate when they are listening.

Furthermore, The Edison Realism Test defines "realism" in an interesting way, as an "emotional reaction" reflecting the listener's earlier emotions when listening to a similar performance. This definition raises the question of what "reality" precisely the recording is meant to reflect, the emotional reality of the performance? If so, why would a recorded voice or instrument necessarily mirror exactly another completely different performance? The image of "His Master's Voice" is most forcefully meant to represent the notion of "fidelity," and fidelity could more or less mean true to the sound of the recorded performance, but the point has been made many times that sound fidelity has always been mediated by the technology, never more so than in the days of acoustical recording, when instruments and voices had to be modified in order to even be "heard" by the recording apparatus. Katz notes that early acoustic recordings replaced strings with brass instruments and that new versions of instruments such as the Stroh violin needed to be invented in order to register in the recording horn (Katz 2010, 44). Kenney says that early recordings feature more men than women, because women's voices were far more difficult to record (Kenney 1999, 42). Furthermore, singers and instrumental soloists recorded at the mercy of "pushers," who were responsible for guiding them into and pulling them out of the recording horn as the pitch and volume of the performance required (Kenney 1999, 41).

Perhaps it would be better to think of "fidelity" at this stage in recording history rather as a category of marketing than as a claim for the sound that could be reproduced. Magoun cites examples of its use in this sense in early gramophone advertising, in a deliberate attempt to counter charges that the gramophone could not reproduce natural sounds sufficiently to be of use to home listeners such as, "blasty, whang-doodle noises are not desired by citizens of culture" (Magoun 2000, 187). Berliner's company countered with testimonials from members of the John Phillips Sousa band and with claims such as, "Its sole purpose is that of Entertainment—Reproducing Everything in Speech or Music, with fidelity to the originals positively marvelous" (Magoun 2000, 188). In fact, early record industry advertising made remarkable claims about Italian tenor Enrico Caruso, arguably the first recording industry star, stating that the recording and the person were inseparable (Suisman 2009, 134)! David Suisman says of such claims that although they are "unremarkable in advertising today, in the 1910s they conveyed an important contention about what sound recordings were and posited a new relation between representation and reality" (2009, 134).

Suisman notes that the human voice carried "particular cultural resonance" at the time of Caruso's popularity, which may itself have contributed to it (Suisman 2009, 140–141). Mark Katz may well point to the disembodied nature of the recorded voice as a "source of great anxiety" at the start of the era of recorded music, quoting an English music critic as saying that some listeners "cannot bear to hear a remarkably life-like human voice issuing from a box. They desire a physical presence. For want of it, the gramophone distresses them" (Katz 2010, 22). Suisman rather highlights the ability of the technology of the time to "convey the grain of the voice," saying of Caruso's records that they "created a semblance of real bodily presence far beyond what images are capable of, an impression produced, if not by Caruso's unique soul, then by the way the sounds he uttered resounded in the muscles and bones of his own unique physical being" (2009, 141). He compares the mediation of the recording apparatus and the way that it altered perception to the close-up of the motion picture. Could it rather be said in this sense that the recording apparatus instead of *reflecting* the fidelity of the "soul" of the human voice actually *created* it? Not marketing hype, but rather, as the listening subject, do I learn what I am listening to in the very act of listening, the mediation of the recording technology contributing to the effect?

Perhaps, then, it might be more helpful to consider "fidelity" to be more in the nature of a promise of an ever closer approximation of an always already deferred "truth," something akin, in fact, to André Bazin's "myth of total cinema,"

> the accomplishment of that which dominated in a more or less vague fashion all the techniques of the mechanical reproduction of reality in the nineteenth century, from photography to the phonograph, namely of an integral realism, a recreation of the world in its own image, an image unburdened by the freedom of interpretation of the artist or the irreversibility of time. (Bazin 1967, 21)

The literary critic George Levine says of the impulse of Victorian realism that this myth could be said to represent that rather than an attempt to reflect its world ontologically, it was rather "a struggle to reconstruct a world out of a world deconstructing" (Levine 1983, 4). The "voice" of "His Master" in this sense might best be thought of as a representation of the attempt to transcend the conditions of its own production. In fact, two facts about the image are elided *by* the image. The first is that the talking machine pictured in the trademark could only reproduce, so the chances of it being the "voice" of Nipper's master would be very slim indeed. The second fact the image elides is related to the first and is actually buried beneath the talking machine in the picture. Berraud originally painted Nipper intently listening to an Edison cylinder machine. Berraud could not interest The Edison Bell Company in his painting, because he was told, "dog's don't listen to phonographs" (EMI Records Ltd. 1997). In other words, the image of the dog at the phonograph horn that presents itself as a representation "fidelity" is itself based on at least three different levels of misrepresentation.

But perhaps this confusion mirrored the confused birth of the technology as well. The ground from which the modern recording industry emerged was actually a battlefield of claims and counterclaims, patents and patent infringements on which nothing was certain. In her discussion of patent and copyright law related to the introduction of recording technology, Lisa Gitelman notes that "recorded sound helped to modulate the already Gordian politics of popular music," pointing to the "otherness" of the dog in Victor's advertising as representing "the new hungry mimesis of the recording phonograph [that] itself came to market larded with assumptions about sameness and difference, about cultural appropriation and assimilation" (Gitelman 1999, 124). The promise that early recording technology held was in fact the promise of the dissolution of the difference between signifier and signified. Richard Osborne notes in this context that the names of both the phonograph and the gramophone "derive from the Greek for 'sound-writing,'" and that it was the unfulfilled hope among the earliest proponents of recording technology, particularly Thomas Edison, to be able to read the language in the groove (Osborne 2012).

While Gitelman makes the salient point that the invention of the phonograph and its separation of sound from sight complicated ideas of mimesis significantly, particularly in terms of the perception of "sounding 'black,'" at the same time she makes the point that "sounding 'black'" was very much considered a possibility in the case of the late nineteenth and early twentieth century rage for recorded "coon songs," derived from the nineteenth-century minstrel show, which she says "played off a contrived sense of authenticity while it also relied on counterfeiting" (Gitelman 1999, 133–136). Gitelman points out the fact that in order to avoid confusion, sheet music publishers "sometimes published minstrel songs with pictures of their blackface performers both in and out of makeup," and she notes that the Buckeye Music Company once took out an advertisement denying that Arthur Collins, a famous performer of coon songs, was Black, as some of his listeners had mistaken him (Gitelman 1999, 136)! "Fidelity" in this sense can only mean being true to the tropes, prejudices, and anxieties of the racial politics of the late nineteenth and early twentieth centuries, "truths," according to Gitelman, that recording technology complicated even as it reinforced them (Gitelman 1999, 135). In other words, "fidelity" expresses the truth of ideology in addition to being an expression of the ideal of the truth of sound.

Whether as myth, promise, or ideal, the notion of "fidelity" has remained remarkably consistent as a driving factor for developers, for marketers, and for individual listeners in the roughly 140 years since the introduction of recording technology, as has the trademark "His Master's Voice," which is now known as one of the most recognizable advertising icons of the twentieth century (EMI Records Ltd. 1997). A Berliner hand crank talking machine advertisement in 1896 states that the

> Gramophone does not imitate, but actually reproduces with lifelike fidelity, purity of tone, distinctness of articulation, all the varying modulations of pitch, quality, and volume of the Human Voice in Speech and Song, the Music of Bands, Orchestras, Solo Instruments of every conceivable kind, in fact, everything within the range of sound. (Berliner Gramophone 2011)

A later advertisement from 1897 talks about a record called "The Morning on the Farm," that "gives a perfect reproduction" of every animal in the barnyard (Berliner Gramophone 2011). A 1917 Edison ad offers the "literal Recreation of the great art of the greatest artists," and quotes a New York newspaper as calling it "the phonograph with a soul" (Music Ads. 2010). A 1926 ad for Columbia Records offered Christmas records recorded using the new electrical recording process, featuring "a new realism never before possible" (Stoddard 2011). A 1949 brochure for the McIntosh AE-2 Equalizer Amplifier Control promised "a performance of unbelievable realism," using bass and treble controls that had been in development for over a year (McIntosh n.d.).

Alexander Magoun makes the point that one early attempt to market the superiority of sound itself was the introduction of the Orthophonic electrical recording system by the Victor Talking Machine Company in 1925. The official introduction was "deferred until inventory had been sold off and the recording staff had built up an electrically based repertoire of discs" (Magoun 2000, 275). While earlier attempts to sell acoustic talking machines were based as much on appearance as on sound, Magoun describes this campaign as a "revolution in the company's advertising" (276). He continues, "In 1925, the novelty was in the sound. An exclamation point after 'Orthophonic' and the repetition of the word 'new' helped persuade potential consumers agree that 'Hearing is necessary because description is inadequate'" (Magoun, 277). In fact, hearing *had to* supplant description in this case, since at this early point in the era of electricity, few consumers knew what the term "electrical" meant or even "understood the power system pervading their lives" (Magoun, 277). A 1927 Victrola advertisement attempts such a description, however, and it sounds not unlike earlier attempts to describe the fidelity of talking machines,

> This amazing instrument brings you vocal music in all its original purity and power. Tones of correct, natural volume; neither too thin nor too loud, but full, round and mellow. The new Orthophonic Victrola captures the very personality of the artist. (Orthophonic Victrola 1927)

As before, the ability of the machine to capture and reproduce the human personality is the marketing promise of fidelity.

"Fidelity" as a notion didn't only survive in the dreams of advertisements and brochures. Jonathan Sterne refers to "fidelity" as "an amazingly fluid term," binding together "[f]unctional, aesthetic, social, and philosophical issues" (Sterne 2003, 216). The history of the technological development of sound recording throughout the years of the twentieth century can be seen as a search forever increasing levels of "high fidelity." The magazine called *High Fidelity* began publishing in 1951 and continued until 1989, well into the digital era, and in that time it covered developments from the introduction of the 45 rpm single and the microgroove long player in the late 1940s through the introduction of magnetic tape, stereophonic records in 1957, and up to the digital compact disc in the 1980s (Riggs 1989). It began as a specialty publication for audio engineers, and it continued, at least at the beginning of its existence, to concern itself mostly with the science of sound, even after it transformed into a periodical aimed at the growing "audiophile" audience. A Jensen ad in the March 1954 edition of *High Fidelity* talks about the "exploitation" of the term by unscrupulous manufacturers of sound equipment, while their use of it denotes "a specialized science and art practiced by too few manufacturers having a history of leadership, progress and intrinsic sense of dollar value" (Quality 1954). At this point, the use of "high fidelity" is so common that some manufacturers feel the need to clarify their intended meaning in using the term. Jensen and other manufacturers of audio equipment advertising in *High Fidelity* back their claims with extensive technical specifications regarding "frequency response," "hum level," and "audio output," among other aspects of sound reproduction (Harmon Kardon Ad 1954).

In 1953, Bogen Sound Systems published a pamphlet called *Understanding High Fidelity,* whose aim was to educate consumers and budding audiophiles about the true nature of high fidelity, a booklet that could be purchased through the mail for twenty-five cents, as published in the pages of *High Fidelity* (Noted With Interest 1954). *Understanding High Fidelity* defines the term "essentially" as "a way of removing every technical barrier between you and what was sung or played on a musical instrument in a recording studio or concert hall" (Biancholli and Bogen 1953). Interestingly, high fidelity sound offers to bring "the listener and composer on more intimate terms of communication. This definition of "high fidelity" promises to erase *all* levels of mediation between composer and listener, not only the recording and the playback apparatuses. In this sense, high fidelity becomes a fulfillment of the late-nineteenth century promise of "non-interventionism," "a growing desire in the world of science to strive for mechanically instilled objectivity in scientific illustrations" developed from the understanding of the camera obscura and the introduction of photography (Slatoff-Burke 2005). *Understanding High Fidelity* calls it "a true test for the recording artist," because "high fidelity can only be as good as the performance; what is interpretatively bad comes through as faithfully as what is interpretatively good" (Biancholli and Bogen 1953, 1).

This definition of "high fidelity" highlights the tension between the objective (score) and the subjective (interpretation) in the way that the recording apparatus responds to the recorded performance, a tension similar to the tension the Bogen pamphlet also points to between the recording apparatus itself and the listener or consumer. In a later discussion of loudspeaker choices, *Understanding High Fidelity* leaves it to the individual listener to decide, saying, "We're not prepared to say which [choice] is the best simply because we do not feel that any person can prescribe here the choices for another. Once a minimum level of quality of manufacture has been established, personal preference is really the only valid criterion left for selection" (Biancholli and Bogen 1953, 38). This appeal to the subjective ear of the individual is taken back almost as soon as it is stated. The next section of the booklet, titled "How to Conduct a Listening Test" states the case directly, saying "It may seem presumptuous to offer suggestions on how to go about listening to equipment after telling you that the subjective factor plays such a large part in selection of components for a high-fidelity home music system" (Biancholli and Bogen 1953, 39). Such advice is necessary, however, because "the human ear is fallible—and notorious for its bad memory" (Biancholli and Bogen 1953, 39). In her discussion of the impact of the camera obscura on the introduction of photography, Megan Slatoff-Burke quotes Daston and Galison on the influence of the machine in nineteenth-century science as saying that the "[p]olicing of subjectivity by the partial application of photographic technology was widespread" (Slatoff-Burke 2005). *Understanding High Fidelity* expands this project into the audio realm, and in the process it narrows our focus of the continuing value of the "myth of 'fidelity'" to the production and the consumption of sound recordings. In addition to being a promise and a marketing tool, "fidelity" gains its own particular valence as a discursive practice.[1]

Discourse is meant in this sense less as a way of talking about sound reproduction and more in the Foucauldian sense of productive of power relations and constitutive of the subject. The notion of fidelity, whether "high" or not, has been remarkably productive both in reproducing discourse communities and expanding power relations over the past century and a half. Industrial, market, and technological forces all may have worked to bring the discursive practices of fidelity into being, but once it emerged, it took on a life of its own, laying the groundwork for the emergence of the listening subject. In her discussion of Foucault and discourse, Sara Mills notes that a "discursive structure can be detected because of the systematicity of the ideas, opinions, concepts, ways of thinking and behaving which are formed within a particular context, and because of the effects of those ways of thinking and behaving" (Mills 1997, 2004). Mills says that these "discursive frameworks demarcate the boundaries within which we can negotiate," in other words, these frames make thinking, speaking, and acting possible. In the case of the fidelity of recorded music, particularly high fidelity, policing the boundaries between subjectivity and objectivity provides just such a framework.

Returning to *High Fidelity* magazine's issue of March 1954, we find a letter of complaint written by Morton Blender, a classical music radio announcer from Providence, Rhode Island. Blender's complaint regards an "epidemic" of poorly tracking LP records of classical music from several different manufacturers, which he likens in his letter to a "disease" (Blender 1954). Blender's complaint also seeks a response from someone in the record industry, and *High Fidelity* forwards his letter to one Remy Van Wyck Farkas, the A&R director at London Records, one of the manufacturers singled out in Blender's letter. Farkas answers with some degree of defensiveness, even as he appears to try to avoid sounding defensive,

> I have spent 23 years in the recording industry and the only thing that I have ascertained that has consistency in it is this. Every time someone has trouble with a gramophone record, the blame is automatically put with on disc—hardly ever on the machine being used to play it! It was the same on 78 RPM, is more common on 33 1/3 RPM and probably was a problem in the days of the cylinder. (Van Wyck Farkas 1954, 23)

After skillfully deflecting the blame in this manner, Farkas continues his response by saying that while manufacturers do a "brilliant" job pressing discs, they have been "completely asleep as to caring what equipment it was played on" (Van Wyck Farkas 1954, 23). Finally, however, Farkas gets to the heart of his response, offering Blender that first, there are no perfect recordings; second, that no disc is ever "NEARLY" as good as the "actual performance;" and third, that while there is equipment capable of playing the best modern records, it is expensive (Van Wyck Farkas 1954, 24).

For the recording industry for whom he presumably speaks, Farkas offers the suggestion that they should "DO INTENSE PROPAGANDA WORK TO SEE THAT HOME EQUIPMENT COMES UP TO THE LEVEL OF THE RECORDING TO BE PLAYED ON IT" (Van Wyck Farkas 1954). In other words, there are no manufacturing defects that are causing the problems outlined by Blender, and if there are, it is the fault of the equipment, or even human error, since Farkas earlier points to the fact that aside from the lack of knowledge by individual consumers even those tasked by manufactures of equipment with demonstrating it to potential customers make bad assumptions about how to set it up to play the records properly. The answer is ultimately in further discourse, since Farkas recommends educating both consumers and equipment manufacturers with "intense propaganda" as to the nature of recordings pressed by the industry. Farkas advocates FFRR or full frequency range recordings, developed by Decca Records in England in the

1940s, and, perhaps not surprisingly, seems to have a problem with the "high fidelity" designation, emphatically stating in capital letters.

> MOST IMPORTANT. STOP THE DAMN NONSENSE ABOUT HIGH FIDELITY AS THERE IS ABSOLUTELY NO SUCH THING. WE ARE IN THE MUSIC BUSINESS AND OUR EFFORTS HAVE ALWAYS BEEN DIRECTED TOWARDS AN INTELLIGENT MINORITY WHO DO NOT NEED OR DESIRE CIGARETTE-TYPE CAMPAIGNS TO ASSURE THEM THAT MOZART, HAYDEN, OR BEETHOVEN CAN SOUND WONDERFUL [sic]. (Van Wyck Farkas 1954, 24-27)

This emphatic denial is more than interesting for being issued in a magazine both titled and dedicated to the idea of high fidelity! The recording industry instead wants to talk about music rather than sound, performances rather than equipment. Of course, in the process of enlarging the public's "musical horizons to an undreamed of degree," Farkas notes that there have to be "growing pains" (Van Wyck Farkas 1954, 27). The schemes of "unscrupulous get-rich-quick characters" will not last, but the music will. Farkas ends his letter by saying that he is planning on leaving the country for a spell after having written it.

"Fidelity" as an idea, again whether "high" or not, grounds both Blender's incredulity and Farkas's defensiveness. Furthermore, Farkas's playful talk of leaving the country after his response is published suggests that he is aware that his discourse, rather than offering the last word on the subject, will in turn generate more discourse. A little over a year later, in fact, *,High Fidelity* carried a letter much more direct and less polite than Blender's had been. In May of 1955, S. Skiner talks about an earlier issue of the magazine carrying complaints about the "shocking quality of LP records" (Skiner 1955, 32). Skiner says that he is "fed up with spending $5.95 for a collection of pops and crackles," and while he shares Farkas's positive attitude about the "enormous strides made in the recording of the performance" by the recording industry, he expresses disappointment with the record itself, calling it "the weakest link in high fidelity at the present time" (Skiner 1955, 33). A discourse community is thus created based on both the promise of a perfect recording medium and the dissatisfaction that it doesn't yet exist. This discourse, by the way, is still going on using many of the same terms in 1979, over twenty years later, as the letter from Pat O'Brien in *Billboard* magazine will attest (O'Brien 1979, 20). In fact, James T. Russell also experienced the dissatisfaction that Blender wrote about in 1954 and O'Brien in 1979, but as an engineer he was equipped to do something about his complaint, which led to the invention of the "red book" level of the compact disc (Knopper 2009).

What both the complaints and the response tend to elide is the myriad levels of mediation between the vaunted performance, itself a level of mediation between composer and auditor, and record that reproduces it on what quickly became known, for better or worse, as a "hi-fi system" at home.[2] In *Discipline and Punish*, Foucault points out the fact that "Prison 'reform' is virtually contemporary with prison itself: it constitutes, as it were, its programme" (Foucault 1979, 234). The movement to "reform" the products of the analog recording industry, as it were, began almost simultaneously to the invention of recording technology, and it was still being widely discussed up until it ended with the birth of the CD. Furthermore, when the analog record began its reascendance in the early years of the twenty-first century, the program of reform as discursive practice reemerged along with it. Along with pressing quality, the issue in the 2010s is whether a vinyl pressing is digitally sourced, meaning that it was mastered from digital files. In his review of a boxed set of Turtles 45s, the website Analog Planet complains, "I say that because I am a Turtles fan and I have their White Whale album *The Turtles! Golden Hits* (WW115) and it's got the analog wet while these singles (some mono, some stereo) have the digital dry heaves" (Fremer 2014). He concludes his review wishing that the makers "would take themselves and their recorded output a bit more seriously" (Fremer 2014).

In fact, the discourse of fidelity has taken a turn in general toward the debate over whether analog or digital sound is "superior." Of course, a large part of the justification for the return to vinyl records is the perception of superiority of analog to digital sound, and there are strong advocates on both sides of the debate. Neil Young, for instance, has waded into the controversy as part of his attempt to sell a new high-resolution digital music system that he calls "Pono." *Billboard* magazine quotes him at length:

> A lot of people that buy vinyl today don't realize that they're listening to CD masters on vinyl and that's because the record companies have figured out that people want vinyl," Young said. "And they're only making CD masters in digital, so all the new products that come out on vinyl are actually CDs on vinyl, which is really nothing but a fashion statement. (Flanagan and Schneider 2015)

The article that quotes Young makes the significant point that "there's a lack of scientific consensus that even the most trained ears are capable of appreciating the kind of hi-res digital audio found on pricey players like Neil Young's Pono" (Flanagan and Schneider 2015). The authors of this article link to an earlier one that quotes Ethan Winer, a high-profile audio engineer as saying the "hi-res movement is based on delusion," which "has been proven time and again with proper blind tests. The problem is that most people don't do proper tests or understand how digital audio works and rely on theoretical arguments instead" (Raile 2015). An earlier NPR interview quotes Scott Metcalfe, director of recording arts and sciences at the Peabody Institute of Johns Hopkins University, on his own preference for the CD format over the vinyl record as saying, "it's—the closer thing to what I'm sending into the recorder is very much what I'm getting back out. With analog formats, although the sound can be very pleasing in certain styles, it's definitely imparting its own sound on it" (Dankosky 2012).

The same claims are often made about the superiority of analog over digital sound. Michael Waehner says in this regard, "A record in good condition played on a reasonably good sound system will rival or surpass a CD under similar conditions, depending on the mastering of both" (Waehner 2012). John Silke says, "play a disc in excellent or mint condition on a quality piece of equipment and it sounds pretty unbeatable" (Silke 2013). Brett Milano points to the fetishistic aspect of the preference for vinyl when he says, "Placing the needle in the groove is a physical act—maybe a sexual one, if you really want to stretch the metaphor—and it's not the same as pressing a button on your CD player, where you can't even see what's going on" (Milano 2003, 2012). The question in the case of both the digital and the analog ideals of fidelity is this: Against what standard are you measuring these types of sound reproduction?

J. Gordon Holt in *Stereophile* calls what audiophiles are chasing a "dream of sublimity," and notes that it can even warp their judgment until it becomes the "Ultimate Truth" affecting even how they come to hear live music performed (Holt 2014). Holt instead imagines a world in which what can be heard in a live performance in a concert hall can be reproduced using a home sound system, and bemoans the fact that supposed audiophiles do not share this desire.

In "Hi-Fi Aesthetics," Joshua Glasgow argues extensively that high fidelity, in the sense of "transparency," meaning one-on-one correspondence between the sound and the recording, is indeed a fact that can be philosophically proven. Glasgow says that hi-fi aesthetics makes the "modest" claim that "it is in principle possible to get a transparent recording of a musical performance" (Glasgow 2007, 173). Glasgow differentiates between "hi-fi," "low-fi," and "no-fi" aesthetics. Lo-fi aesthetics holds that "while recordings might be able to capture some elements of the music being recorded, certain distortions are inevitable" (Glasgow 2007, 163). No-fi aesthetics, on the other hand, "holds that it is in principle impossible to capture even a partially faithful, transparent recording of a musical performance" (Glasgow 2007). Glasgow argues that the terms of the debate over the possibility of the transparency of recorded music have mostly been set by the lo-fi and no-fi sides of the debate, but he claims that the hi-fi position is more modest, and that the "hi-fi ideal, then, is to get the perfectly transparent recording" (Glasgow 2007, 164). The greatest obstacle to fidelity, however, is the very fact of the record itself. In order to be true to the note that it "recreates," the recorded note would have to decay and die; the recorded note can be repeated forever, as long as the technology exists to replay the recording. The endless repeatability of the recording tells against it. What is important for this discussion, however, is the discursive power of this search for transparency.

Of the debate between digital and analog sound, Glasgow says "What matters is what is presupposed in such debates, namely that there is a *better*" (Glasgow 2007, 164). Again, the line between the subjective and the objective comes into play. Silke speaks for many vinyl lovers when he says that in comparison to the CD, the "LP [is] warmer and more real sounding, the CD [has] a tinny unpleasantness to it" (Silke 2013). Sean Olive, the director of acoustic research at Harmon International, states that there are

> a number of factors involved in our perception of sound quality, and a lot of them have nothing to do with the sound itself. So in the research that we do, we've looked at the—and we call these things nuisance variables or biases. So one of the ones that we deal with is psychological nuisance variables, which have to do with your knowledge or expectation of what you're hearing. Thomas Edison knew this a hundred years ago. He said people will hear what you tell them to hear. (Dankosky 2012)

Joshua Glasgow believes that the history of fidelity has been a clear trajectory of technological change leading to greater transparency. He argues,

> After all, is it not obvious that we have gotten closer to true hi-fi? Surely stereo is more transparent than mono, for instance. And capturing the performance faithfully is undoubtedly one skill sought by aspiring recording engineers, even by those who go on to intentionally distort the performance for the sake of art (producing a unique kind of art, different from the merely documentary recording, variously called "works of phonography," "recording-artefacts," and "art phonography"). (Glasgow 2007)

Alexander Magoun, however, makes the point that the through line of technological development was by no means a clear one, and in fact he highlights all of the disjunctions and compromises that play a part in the development of recording technology on its bumpy road to high fidelity. Magoun says, "Technological innovation continued to be only one aspect of business strategy. Corporate entrepreneurs had to find ways to sell a new sound in the 1920s, to sell records at all in the Depression, and to exploit the best market in industry history during the 1940s—before, during, and after a world war" (Magoun 2000, 216). In fact, most of what might be termed the technological advances of the sound recording industry were themselves more a response to market forces, such as the introduction of radio, the Great Depression and with it the collapse of the recording industry, and the world wars, than they were of the desire for technological improvement of existing technologies (Magoun 2000).

For instance, according to Richard Osborne, the triumph in Britain of the technologically inferior flat disc over the cylinder in the first of the recording industry's many format wars had more to do with the advertising superiority and greater repertoire of the Victor Talking Machine Company (Osborne 2012). Marketing savvy, however, was not the only aspect of the industry that influenced technological development. The introduction of higher quality materials and technologies required for a genuine "high fidelity" product were significantly delayed by the Depression and then the war. Magoun says that while the sound of radio was significantly improved, the phonograph, which "contributed less than four percent of RCA's revenues in mid-decade received far less priority for research than other, purely electronic technologies" (Magoun 2000, 300–301). Furthermore, Magoun argues that the "expansion of 'high fidelity' technology, an industry to produce it, and an audience to appreciate it took place in large part as a consequence of World War II" (Magoun 2000, 444). The war also helped to create the technologies, such as magnetic recording, that made the expansion of recorded frequencies possible, and it also "raised the consciousness of a whole generation of men to the possibility of high fidelity, and gave them the skills and equipment to develop it" (Magoun 2000, 427). Even the format war between the 33 1/3 LP and the 45 at the end of the 1940s and the triumph of the LP as conveyer of culture were influenced as much by market forces as technological ones (Magoun 2000, 515–519).

In his study, *Capturing Sound: How Technology has Changed Music*, according to Mark Katz, just as digital technology has gotten us presumably to the brink of the transparency for which Glasgow argues, the introduction of the MP3 has brought us to the "age of post-fidelity," in which "it is convenience, not quality that is paramount to most listeners" (Katz 2010, 217). Katz argues that for most of the history of the recording industry, what he calls "the discourse of realism" influenced "the way people have conceived, characterized, and valued the technology of recording since its inception" (217). Katz's formulation of the discourse of realism as "the idea of recorded sound as the mirror of sonic reality" roughly corresponds to the discourse of fidelity that is the concern of the current discussion (Katz 2010). The discourse of fidelity, however, lives on in several ways, not the least of which is the resurgence of the inconvenient vinyl record in the digital era. First, the introduction of Young's Pono system and "high resolution" formats in general demonstrates that there is still at least a market for the kind of fidelity Katz thinks has been forsaken. Second, low-bit-rate digital music may have been the standard in the age of Napster that Katz talks about, but it is no longer the standard, which means that there must have been a consumer demand that drove the change. The largest retailer of digital music, iTunes, may have started out offering digital sound files at 128 kbps, by 2007 it was offering iTunes Plus files at 256 kbps (iTunes Plus 2007), and as of 2012, they introduced a new high fidelity standard that they called "Mastered for iTunes," which if it isn't an alternative to high resolution audio, sounds much better than regular sound files according to the reviewers (Lendino 2012). In fact, the "Mastered for iTunes" page calls them "virtually indistinguishable from the original master recordings" (Learn More n.d.). Third, even when you are talking about listeners in the "age of post-fidelity," you are still using the term "fidelity" as a standard.

What is important is the discussion itself, which has captured imaginations, driven technological advances, and generated both communities and industries. As in Bazin's myth of total cinema, if the one-to-one correspondence between the recording and the sound recorded were ever to be realized, the myth itself would cease to be, which in his discussion of the *Bicycle Thief*, he refers to as "No more actors, no more story, no more sets, which is to say that in the perfect aesthetic illusion of reality there is no more cinema" (1971, 2005, 60). The object of fidelity remains, as it must, always already deferred. Going all the way back to *Understanding High Fidelity*, the discussion of fidelity almost always includes diagrams of waveforms and graphs representing frequency response rates, and the evaluation of any individual piece of equipment or technology in magazines such as *High Fidelity* also centers on such objective measures as representations of sound quality. At the same time, these discourses almost reflexively return to the notion of the subjective as the determinant of individual perception. Returning to Bazin for a moment more, what he says about realism in relation to the cinema helps to clarify this position; he says that "realism can only occupy in art a dialectical position—it is more a reaction than a truth" (1971, 2005, 48). In the case of reproduced sound, it might be useful to ask, a reaction to what?

Despite all of the objective measurements the discourse of fidelity throws up, there is no escaping the fact that, as *Understanding High Fidelity* calls it in 1953, "the subjective factor" always was and still is a significant part of the listening experience (Biancholli and Bogen 1953, 39). David Steffen makes this point when he says, "it's worth remembering that the music-listening experience is always biased. Each of us listens to music and hears out through a filter that is shaped, much as our entire emotional and intellectual character is shaped, through the ever-present nature and nurture: what we are given as part of ourselves and what we acquire as our lives progress" (2005, 10). In an article about the introduction of the Pono music player, Stephen Shankland cites a 2007 Boston Audio Society experiment in which listeners did no better than "random guessing" at knowing the difference between high resolution audio and normal resolution audio, neither the first nor the last study to yield such a result (2014). And, it isn't only the supposed superiority of high-resolution audio that is in question. Of the superior sound of vinyl, Chris Kornelis quotes Jim Anderson, a recording engineer of note and a professor at NYU's Clive Davis Institute of Recorded Music as saying, "I think some people interpret the lack of top end [on vinyl] and interpret an analog type of distortion as warmth . . . It's a misinterpretation of it. But if they like it, they like it. That's fine" (Kornelis 2015). Mark Richardson describes this misinterpretation in more detail,

> The "warmth" that many people associate with LPs can generally be described as a bass sound that is less accurate. Reproducing bass on vinyl is a serious engineering challenge, but the upshot is that there's a lot of filtering and signal processing happening to make the bass on vinyl work. You take some of this signal processing, add additional vibrations and distortions generated by a poorly manufactured turntable, and you end up with bass that sounds "warmer" than a CD, maybe—but also very different than what the artists were hearing in the control room. (Richardson 2013)

In other words, our desire for unmediated sound cannot help but come up against the exigencies of the mediation of the machines that reproduce them for us, and what we are hearing becomes a compromise between the two. The question is how we get to the point where we accept distortion as truth. The discursive practices that grow out of this tension contribute to the constitution of listening subjects, whose continued demands for ever closer verisimilitude can both lead to the introduction of new technologies such as the CD, the SACD, and the high-res FLAC file, *and* the recovery of old ones, such as the analog vinyl record.

To enter into the discourse of fidelity requires training. It also requires "investment," both economic and aesthetic, on the part of the listener, as Eric Rawson puts it in his study of audiophilia, "Perfect Listening" (Rawson 2009). Of the twenty-first century audiophile, Rawson says,

> Contemporary high-end listening likewise requires a great deal of training in the form of advertising, articles, editorials, salespeople, conventions, exhibitions, testimonials, and graphic representations, not to mention the more general educational contexts in which one acquires cultural capital, before the listener accepts his experience as a natural and meaningful one. (Rawson 2009, 205)

Of course, the case can be made that this has been the case since the very beginning of sound reproduction. Jonathan Sterne notes that "[d]uring the early history of commercially available sound reproduction technologies ... [p]eople had to learn how to understand the relations between sounds made by people and sounds made by machines" (2003, 216). Saying a recorded sound is "natural sounding" is already a radical redefinition of "nature." In fact, Rawson says that the listener's embrace of the artificiality of both sound and reproduction is an important part of the listening experience. The mediation of both the equipment and the listener's expectations are required at the same time they are disavowed as determinants in the listening experience. The listener "must willfully ignore his conscious role in the process; he can know the music only by denying that this knowledge is fictitious" (Rawson 2009, 209).

But, what is it we are being trained to do? The discourse of fidelity provides a frame for speaking, thinking, and acting in relation to the technologies of sound reproduction that together contribute to the constitution of the listening subject. Rawson separates audiophiles from lo-fi and no-fi listeners in his taxonomy, saying that audiophiles through their "embrace of the ambiguous" can "exercise the greatest degree of agency" (Rawson 2009, 210). But the question remains, agency to do what? Among the things that the early discourse on fidelity, including Nipper and "His Master's Voice," taught listeners was the very idea that recorded music is a thing, what David Suisman refers to as the "transformation of American musical culture" (2009, 10). As a result, according to Suisman,

> Consumers assimilated the idea of music as issuing from an automatic machine (such as a phonograph or player piano), detached from human labor, and fixed in objects (such as records or piano rolls), portable and storable, independent of time and place. Music, which had once been produced in the home, by hand, was now something to be purchased, like a newspaper or ready-to-wear dress. (2009, 10-11)[3]

Along with sound being sold to the listening subject as an object, the illusion of control over these objects was also sold as an important property of subjectivity. William Kenney writes, "With a phonograph purchase, the customer bought a form of individual control over the cultivated, refined, and complex world of music, without controlling who or what was recorded, where, when, or how" (1999, 53).

As Rawson points out, that same subject/object ambiguity extends to the listening subject's relationship to reproduced sound. There are two poles of the listening experience in the image of "His Master's Voice," Nipper and the voice. Projecting yourself into that image inevitably raises the question of with whom you are supposed to identify. Are you the dog, in the listening pose of recognition? Are you the master, certain that your voice will live on after your death? The discourse of fidelity places the listening subject on the line balancing precariously (ambiguously?) between the two. The debates over fidelity all center on the same question, does the sound originate with the recording, or with the listener? According to Chris Kornelis, the aforementioned Jim Anderson teaches a course in how to recognize the difference between normal- and high-resolution sound, quoting Anderson as asserting, "Someone has to say to you: Listen for this, listen for this, listen for this. And when you start to home in on those details, it starts to become very clear." (Kornelis 2015). If you need to be taught how to listen in this way, are you also being taught that what you are currently hearing you are also mishearing? Anderson appears to be teaching listening subjects to recognize the sounds they are hearing as outside of themselves. Pedagogically speaking, then, we are still in the land of the dog, learning to (mis)recognize the voice we are hearing as not originating in our own heads.

NOTES

1. See Jonathan Sterne in particular for a discussion of the notion of fidelity as a discursive practice: "The history of sound fidelity, as an operative concept, a technical principle, and an aesthetic, is a history of beliefs in and about sound production as well as a history of the apparatuses themselves" (2003, 221).

2. In his study "Perfect Listening: Audiophilia, Ambiguity, and the Reduction of the Arbitrary," Eric Rawson describes some of these as "the relationship among the recording artists, the engineers, the producers, the composers, the marketers, the critics, the recording and playback technology, the conditions of composition, the site at which the recording was produced, the site(s) at which it is reproduced, and the audiophile listener's unique physical and psychic condition" (2009).

3. Sterne argues that the general public embraced sound technologies as quickly as they did because they were "artifacts of vast transformations in the fundamental nature of sound, the human ear, the faculty of hearing, and practices of listening that occurred over the long nineteenth century" (2003, 3).

Chapter Two

The Beatles on itunes and Vinyl Reissue

Aesthetic Discourse and the Listening Subject

Jason, a friendly-looking man with fringed hair, a checked shirt, and glasses, is clearly excited as he says hello to the YouTube Vinyl Community at the beginning of his video.[1] "I'm here for another video, and it's a pretty special one," he starts. He continues, "I got into vinyl because I really wanted to have a strong original collection of the records that I love." Holding up a record in its inner sleeve covered in a poly vinyl bag, he says, "One that I've been chasing for a very long time is this," leaving a moment of suspense before he turns the record around to show the front cover, "the *first* pressing of Beatles *Please Please Me*" (emphasis on the word "first"). Jason says that he has wanted one of these records for a very long time. He says that somebody told him once "every so often, one of these records slips through the net," and, he continues, "that's what's happened here," and that he "got it for a very, very good price." He notes that he bought it on eBay, and that "there were no pictures of the disc" included in the auction posting, but that looking at the picture of the cover he "saw the telltale sign which is the 'Angus McBean' is in line with the 'S' and not the 'G' which it is in later pressings" at the bottom of the cover photo below the printed text, "and 12 other songs."[2]

As he pulls the disc out of the cover, Jason says that while it is "not in the best condition in the world, definitely not, but, again, for the price it's amazing." He pulls the disc out "very slowly," and points out "the lovely gold label" that it has "the Dick James credits, so it's *that* one (emphasis his)." After a few words about the condition of the disc, which plays very well despite some marks on the vinyl, and the sleeve, which is "a little bit worn, but it's a first pressing sleeve, so you can't mess with that." Jason concludes that he is "very, very happy," because he has "almost completed" his Beatles collection. His video description says, "this is a proud day," and the title of the video is "Beatles Vinyl Update—Fantastic Grail." Chapter 7 will discuss the YouTube Vinyl Community in more depth and chapter 3 deals with the details of a first pressing of *Please Please Me* as part of a discussion about the search for authenticity in first pressings. This chapter will look at the notion of a vinyl record as a "grail," and consider what a record collector both is seeking and finds in one, using the Beatles in order to discuss aesthetics, especially Benjamin's notion of "aura" in a work of art, and how it attends in the case of such a mass-produced commodity.

As far as the "culture industry" is concerned, "[t]he machine is rotating on the spot," according to Theodor Adorno. How? "Unending sameness also governs the relationship to the past. What is new in the phase of mass culture compared to that of late liberalism is the exclusion of the new" (Adorno 2002, 106). The reproducibility of aesthetic objects leads mostly to just reproduction of the same, in Adorno's formulation. Walter Benjamin also has something to say about reproducibility, noting, "[e]ven the most perfect reproduction of a work of art is lacking in one element: its presence in time and space, its unique existence at the place where it happens to be. This unique existence of the work of art determined the history to which it was subject throughout the time of its existence" (Benjamin 1968, 218). On the subject of "sameness" and its relationship to timelessness, back in 1964, who could have guessed that well into the second decade of the twenty-first century, the Beatles, who haven't even been together as a working concern for forty-five years, would still mean something? *What* they mean, of course, is a question open to debate; *that* they mean, however, goes without question. For instance, according to Kenneth Womack, "the Beatles qualify as a textual entity—not merely for their status as a cultural phenomenon for the ages, but also for their highly-evolved narratological tendencies" (Womack 2007, 165). How, though, does the tendency toward narrative reveal itself? Is it in the songs the Beatles wrote and recorded? Perhaps it is in the recorded albums and singles through which those songs revealed themselves? And, what about the overall arc of their career? Beyond question, perhaps, and in opposition to Adorno's point about popular culture continually revealing through itself more of the same, the cannon of the Beatles continues to reveal itself as "something new," and in its packagings and repackagings, it provides a sort of validating function for whatever sound technologies are used to reveal it. As David Stubbs says in his review of the 2014 box set reissue of their first ten albums in mono, "Even if you've heard the songs a million times, hearing them in mono, (especially on a modern stereo system—the best of both worlds) reveals the familiar fresh" (Stubbs 2014).

That the Beatles are/were an analog phenomenon as a recording entity goes without question. That they later became a digital phenomenon and then returned to analog is perhaps more worthy of discussion. That the Beatles' transition from analog to digital and back again is somehow deserving of extensive media comment is even more significant. "Sound technologies," Jonathan Sterne says, "emerged from a changing context of research, innovation, and development; they grew in the spaces of a transforming middle class, and they were nourished with surplus materials and labor generated by industrial capitalism" (2003, 183). Bartmanski and Woodward add, "[i]t is hardly a random occurrence that giant pop legends of our time like Sinatra, Miles Davis, the Beatles, the Rolling Stones, or the artists of Motown emerged and inspired incredibly intense response at the time when records entered the market as socially available modern products" (Bartmanski and Woodward 2015, 6). This chapter concerns itself with the Beatles as part of the changing context of sound technologies and considers the parts in which their own development as artists, their emergence as a cultural phenomenon play in the innovation and development of sound technologies. How does their journey from analog to digital and back to analog affect our understanding of the ways in which their cannon signifies culturally, and how does their cannon and their history affect our understanding of the sound technologies through which it was disseminated? This chapter seeks to consider these questions, particularly as they pertain to the remastering and reissue of their canonical albums in both stereo and mono in the vinyl format. Do the Beatles in digital or the Beatles on vinyl matter, and if so, in what way(s)?

The Beatles' cannon, if it can be called such, was arguably not fully established worldwide until they entered the digital age with the reissue of their original British albums in 1987. Prior to that, their recorded works were distributed all around the world in many different configurations, some of them, such as in the case of cassette and 8-track tapes, format dependent. In 1987, all the albums were standardized to the British releases, with the exception of the USA LP edition of *Magical Mystery Tour*, which was originally released in the UK as a set of two 45 EPs. Even the twentieth anniversary of the CD reissues was worthy of comment. Martin Belam said at the time, in

> recent months there has been a lot of focus on the digital availability, or lack of, of the Beatles back catalogue. There has been endless speculation that the settling of lawsuits between Apple the computer company and Apple the Beatles music company might herald the iTunes exclusive availability of the Beatles music. EMI's announcement of DRM-free downloads also prompted speculation that this was in preparation for premium Beatles downloads being made available. (Belam 2007)

Belam echoes a common complaint as the years without further reissue passed by, when he says that the "original Beatles CDs featured no bonus tracks, no alternate mixes, and no liner notes, unless liner notes had featured on the original vinyl sleeve" (Belam 2007). On the occasion of the 2009 remasters, the *New York Times* notes the complaints that Belam is talking about, saying,

> In 1987 the elation of finally getting the group's classic recordings on CD, four years after the format was introduced, quickly gave way to disappointment with the discs' sound quality and presentation. Like many early CDs, several (though not all) of the Beatles' discs had a harsh upper range. And except for "Sgt. Pepper's Lonely Hearts Club Band," which was put in a deluxe package with liner essays and archival photos, the 1987 CDs came with minimal, slapdash artwork. (Kozinn 2009)

And, the article also calls attention to the fact that while the tapes used for the remasters are "the same tapes EMI used in 1987, but analog-to-digital technology has improved considerably since then, making it possible to get a much more fine-grained, high-resolution digital transfer" (Kozinn 2009). These releases also marked the digital debut of the first four albums, *Please Please Me*, *With the Beatles*, *A Hard Day's Night*, and *Beatles for Sale*, in stereo, having been released only in anemic mono mixes back in 1987. All of the remasters, it was reported, had reverted to the original 1960s master tapes, with the exception of *Rubber Soul* and *Help!*, which continued to use George Martin's remixes for the 1987 reissues (Kozinn 2009). While the stereo mixes became the standard issues for those who wanted to purchase the individual CDs, the mono mixes of all albums up to and including *The Beatles (The White Album)* were made available as part of an expensive boxed set that also included the original stereo mixes of *Rubber Soul* and *Help!*, perhaps indicating the specialized interest—and disposable income—of the fans who might wish to commit to it.

The next significant event in the history of the Beatles' recorded music occurred on November 16, 2010, when Apple began to offer the recorded catalog of the Beatles for exclusive digital download in its iTunes store. This development was reported enthusiastically in every major media outlet, print, broadcast, and online, as a significant benchmark in the history of digital distribution. The Beatles were seen as one of the last acts of their stature not offering their work online, a fact that carried its own weight in the music industry. In a blog post, the *New York Times* reported on Apple's teasing announcement on their website the day before, in such a way that it highlights the significance with which the music business at least saw the next day's announcement, including a screenshot of the Apple home page, and quoting the text,

> Tomorrow is just another day.
> That you'll never forget.
> Check back here tomorrow for an exciting announcement from iTunes.
> (Bilton 2010)

Not many other acts in the world of music would be able to justify such mystery, such a PR push by Apple and by their record label. An article published on the same day solved the mystery, and speculated on its import, saying that because they had been one of "the last major holdouts against selling its music via digital downloads, the Beatles are the ultimate prize for any music company" (Sisario and Helft 2010). Why were the Beatles considered to be such a prize? Womack quotes Ian McDonald about the Beatles' cultural significance, saying that McDonald "notes that 'so obviously dazzling was the Beatles' achievement that few have questioned it.' Their recordings, he adds, comprise 'not only an outstanding repository of popular art but a cultural document of permanent significance'" (2007, 166).

After the release of the Beatles' albums on digital download, what could possibly have come next in their recorded history? Probably no one in the cultural mainstream in 2012 would have guessed that it would be the reissue of all of their stereo albums on vinyl using the 2009 digital remasters, but that was precisely what occurred. Again, the *New York Times* reported this development, saying, "The LPs will be pressed on audiophile quality vinyl, using a version of the cleaned-up masters prepared for the 2009 CDs, with the exception of 'Help!' and 'Rubber Soul,' which will use the remixed versions that George Martin, the group's producer, oversaw in 1986" (Kozinn 2012). The article also notes a potential reason for this development, saying of CDs and digital downloads, "many fans of the group have argued that such newfangled ways of hearing the band are inauthentic—that the experience just isn't the same on anything but vinyl, the format on which the albums were originally released" (Kozinn 2012). Related to this point, the article points to the potential for controversy about the vinyl reissues, saying that the decision of EMI, the Beatles' label,

> to use the digital masters may be contentious among collectors, who have been arguing for years that EMI should release a high-quality vinyl set made directly from the original analog masters prepared in the 1960s. The 2009 masters are digital copies of those recordings, and although they embody a few notable fixes—the repair of a dropout in the guitar part of "Day Tripper," for example—they have been sent through a digitizing process that, some listeners feel, degrades the original analog sound. (Kozinn 2012)

"Fans" of the Beatles, according to this point of view, consider their music to be authentic only in its original analog form. Attempts of the recording engineers to alter and "improve" the sound of the original recordings using digital means is considered to be a degradation of the music and the sound.

Marc Hogan in *Spin* magazine recounts a scene from a film in order to make an aesthetic point that the viewer is supposed to take as in some way given when he says,

> In 1996 Sean Connery thriller The Rock, Nicolas Cage's character spends $600 on a Beatles LP. When asked by a co-worker why he didn't just spend $13 on the then-popular medium known as compact disc, Cage responds presciently. "Well, first of all, it's because I'm a Beatlemaniac," he offers, deadpan as always. "And second, these sound better." (Hogan 2012)

More remarkable than the twenty-first century reissue of the Beatles' albums on vinyl in the first place is the 2014 release of all of the albums up to *The Beatles* in mono. If vinyl is the best way to listen to the Beatles, then perhaps the best vinyl to listen to them on is the monophonic. Shortly after the release of the mono remasters, *Billboard* noted, "[t]he Beatles take over half of the top 10 on the Vinyl Albums chart, as the reissue of the group's mono catalog on LP prompts a wave of Beatle mania on the charts" (Caulfield 2014). The 2012 stereo vinyl remasters had been issued before Apple Records and the Beatles' catalog was transferred in the sale of the "venerable" EMI Corporation to the Universal Music Group (UMG), and they were widely considered to be a disappointment (Sisario 2011). The post-UMG reissue of the mono recordings was widely believed to be a resounding success. Pat Gilbert tells listeners what to expect when he says, "they've issued the recordings on 180 gram vinyl, this time remastered from the original 1/4-inch mono mixes in Abbey Road's tape library" (Gilbert 2014). John Haley says about the mono masters on vinyl that "[t]he sound quality is consistently more lively and "present" than the 2010 'Beatles In Mono' CD box set" (2015, 152). The reason Haley gives for the liveliness and presence is that "[i]n comparison to these new LPs, there was a general 'scooped out' quality to the mono CDs' sound. The CDs sounded somewhat tinny compared to the LPs, but not so much 'worse' as 'wimpy' and 'weaker'" (2015, 152). Clearly, for some listeners at least, the answer to the question of what the Beatles mean, at least prior to 1969, contains the words "mono" and "vinyl." Gilbert quotes the second engineer for *Sgt. Pepper's Lonely Hearts Club Band* Richard Lush as saying, "the only real version of [*Sgt. Pepper*] is the mono version" (Gilbert, Back to Mono 2014). Haley calls the market for mono Beatles recordings on vinyl "a niche within a niche," but that such a market was seen to exist in 2014 is certainly worthy of examination.

Of course, the epistemological connection of the Beatles with the vinyl recordings that bear their name did not begin with them. The story of Caruso's connection with the sound technology through which his voice was disseminated perhaps provides a useful analogue to the Beatles in analog. The first big recording star is widely believed to be Enrico Caruso, who David Suisman calls "a new kind of celebrity, launched by the music industry but not altogether controlled by it, and echoing deeply throughout twentieth-century American and global culture. Caruso both embodied and disembodied the aesthetic and technological possibilities of the phonograph" (2009, 126). Suisman says, "what people heard when they listened to Caruso was never something so simple as a strong and beautiful voice" (2009, 142). Aside from his talent and abilities as a singer, Caruso's meaning as part of the celebrity culture of the beginning of the twentieth century had much to do with his association with the technology of the gramophone and the Victor Talking Machine Company's Red Seal record label. The experience of listening to Caruso at the time of his greatest fame

> was the experience of listening simultaneously to Caruso's singing and to the technology itself, for the two could not be separated—here was the paradox of transcendent vocalization anchored in a mundane shellac disc, the human utterance inextricable from the surface noise of the needle in the grooves, the echo of the horn, the whirring ring and clicking of the machine. (Suisman 2009, 141)

Furthermore, Caruso shares with the Beatles continuing demand for his recordings up to the present day. Suisman says, "over the years his records have been reissued, repackaged, remixed, and re-recorded countless times" (2009, 148). A YouTube search of his name reveals 84,400 results. Perhaps more remarkably, Caruso also has a presence on Spotify, and digital files of his music can be purchased via iTunes.

Caruso also shares with the Beatles the fact that the recording industry has seen fit to alter their original recordings in significant ways, sometimes to the consternation of their fans. The original acoustic recordings of Caruso have been altered to make them sound more "modern" more than once, most recently on the digitally recorded Caruso 2000, which matched his isolated voice with high-fidelity studio orchestra recordings. Suisman says,

> the original recordings were even digitally stripped of their original thin-sounding accompaniments and re-recorded with a live modern symphony orchestra. A skeptical New York Times reviewer condemned the project as "woefully misbegotten" and wondered why the record company would not allow the "great recordings [to] speak for themselves, on their own terms." (2009, 148)

The reviewer to whom Suisman refers had much more than that to say about this project. He calls the project perhaps "an eloborate practical joke," a point made by "the untouched version of 'Vesti la giubba' may be the key to the prank: it's as if the listener, having sat wide-eyed and mouth agape through 16 sonically colorized tracks, were at last allowed to hear Caruso in his own element and to realize that RCA and Mr. Werba were only kidding" (Kozinn 2000). After making the point that the record label altered the recordings in order to make them palatable for modern listeners, a project at which all concerned failed in his opinion, Kozinn imagines other possibilities for the record company to make money from the bastardization of Caruso's recordings before concluding, "[t]he alternative, of course, would be for RCA to let its great recordings speak for themselves, on their own terms, as they always have" (2000).[3] What those recordings might actually have to say were they to speak for themselves is a question that Kozinn leaves unanswered, as he must, but it, and the desire to hear the original recordings speak, and what they might have to say, are questions worth pondering.

The Beatles have, of course, suffered similar mortifications on the corpus of their works, even when they were recording at the height of their career, and certainly after they broke up in 1970. For every critically lauded *1962–1966* and *1967–1970* there is a critically drubbed *Love Songs* and *Live at the Hollywood Bowl*. One such example is "The Beatles' Movie Medley" released in 1982 on Capitol Records in the USA and Parlophone Records in the UK. A Beatles blog notes of this recording, "Capitol Records first issued the single in conjunction with the album *Reel Music* and was inspired by the success of the 'Stars On 45 Medley,' a recording which included numerous Beatles songs sung by a John Lennon soundalike [*sic*]" (Stormo 2010). The album it was designed to promote, a compilation of music from the Beatles' films called *Reel Music*, is referred to by *All Music* as, "a pretty pathetic venture, and clear evidence that Capitol Records and its parent company, EMI, had run out of ideas when it came to marketing the group's music" (Eder n.d.). Even while they were a going concern, however, Capitol Records in the USA was submitting their recordings to similar indignities. The running order of their albums was interfered with, as Capitol cut the albums apart in order to make more albums and combined them with single releases that the Beatles had never intended to be issued in LP format. Furthermore, in a process since referred to disparagingly as "Dexterization," after the Capitol Records executive responsible for the Beatles' USA releases, the original monophonic recordings were reprocessed for Duophonic stereo. Because Dexter found the Beatles' UK recordings to be "thin and dull," he decided, "to boost the high end and, in many instances, add reverb to make the songs sound more cavernous and dramatic. The reverb was especially useful on tracks that were electronically processed for stereo: the wash of ambience helped fill the midfield between the left and right channels" (The Story Behind the Beatles U.S. Albums 2013). When discussing creating stereo mixes of songs for the US compilation *Hey Jude*, Geoff Emerick says, "I was glad that the record company was giving us the opportunity to run off new mixes from the original multitrack tapes instead of subjecting them to the pseudo-stereo processing that had marred so many early Capitol releases" (Emerick and Massey 2006, 313). In order for the meaning of the Beatles' original recordings to be betrayed in the ways fans and critics suggest they have been, however, there needs to be the entity "the Beatles original recordings" to be betrayed. The question of origin seems to be very real, not just in the desire for the taped performances that made up the sources for the Beatles' recordings, but also in the experience of the recordings themselves, which might go some way in explaining the late release of the mono vinyl reissues. Mono was the standard for listening to popular music in the 1960s, or, as Gilbert says, "youth music was heavily mono-minded" (Gilbert, Back to Mono 2014).

Just as in Caruso's case, what the Beatles' recordings might have to say were they able to speak for themselves is also very much an open question. That does not mean, however, that the answer is not being sought. The desire to seek the truth of the Beatles in their original recordings and the fervor with which such desire is expressed are both actualized in the discourse of both critics and fans of the band. In the perhaps parallel case of literature, Mark Currie says this desire for "the successful transmission of truth in discourse is a prelapsarian ideal" (Currie 1994, 360). In this formulation, Derrida "links the Adamic myth to the philosophy of language through the idea of the Fall from presence" which is defined as,

> The Fall from God's presence brought about by Adam's sin is related, for Derrida, to the Fall from presence of the meaning of the sign, which, as a representation, stands in place of a presence which preceded it. The thematics of the sign and the Adamic myth are linked by a common metaphysical supposition which leads to a nostalgic desire for original presence. (Currie 1994, 355)

Currie quotes Derrida's *Of Grammatology* as saying "a possibility produces that to which it is said to be added on," which Currie glosses as meaning, "the history which follows from an origin is seen to inhere in the origin from which it is supposed to have fallen" (Currie 1994, 355). Rob Fitzpatrick makes just such a prelapsarian move when he says, "the truth of the Beatles' story, the reason why, after a million and one retellings, it's as gripping and as startling as it ever was, is that every single move they made was the end of an era. Or, more pertinently, the end of their own era" (Fitzpatrick 2013). Of the later years of their career, Ian MacDonald notes, "it seemed to many fans of the Beatles that the group was somehow above and beyond the ordinary world: ahead of the game and orchestrating things. So in tune were they with the spirit of the times that they almost seemed godlike" (MacDonald 2012, 9). Nick Farmer marked the sense of occasion brought on by the impending anniversary of the Beatles' first album, noting "By the time the Olympics have finished (the next big mark on our calendar) it will have been almost 50 years since the Beatles stepped into Abbey Road Studio No 2 to record their first album (another mark on our calendars). What milestones did it place in music history and how important is it almost half a century on" (Farmer 2012, 74)?

And, in the discussion of the Beatles and the reaction to the digital and the vinyl reissues of their work, the difficult historiographical question of "origin" becomes an aesthetic question of "aura." Mark Katz lays out the case of Walter Benjamin's conception of aura in "The Work of Art in the Age of Mechanical Reproduction," saying,

> While the visual arts concerned Benjamin most, his ideas are relevant here. "Even the most perfect reproduction of a work of art," he maintained, "is lacking in one element: its presence in time and space, its unique existence at the place where it happens to be." Reproductions, therefore, lack what Benjamin called the "aura" of the artwork. From Benjamin's standpoint this absence is to be lamented. He speaks of the withering of the aura, the depreciation of the artwork, the loss of authenticity, and the shattering of tradition. (Katz 2010, 14)

Katz notes the limitations of Benjamin's theory of aura, saying he "missed half of the equation. True, mass-reproduced art does lack temporal and physical uniqueness, yet reproductions, no longer bound to the circumstances of their creation, may encourage new experiences and generate new traditions, wherever they happen to be" (Katz 2010, 15). This point will be considered later in the current chapter, and indeed in later chapters as well. Chapter 3 will also discuss the discourse of authenticity as it particularly relates to the recordings of the Beatles. For now, it might be worth noting how much the question of aura plays into aesthetic assessments of recording entities, and in fact the entire idea of who the Beatles are as a cultural phenomenon. To draw a literary parallel, the question of exactly what I am talking about when I talk about "the music of the Beatles" is no less complex or difficult to decipher with precision than the question of what I might mean by "the works of Shakespeare." Both questions are tied in Gordian Knots by notions of authorship, authority, authenticity, and authorization, notions further complicated in the case of "the Beatles" by the central role played both by technology and by corporate marketing in the creation of recorded artifacts.

Recorded music requires such a complex nexus of electronic technologies to make the sounds, capture them on groove, tape, or hard drive, and turn them into saleable commodities that it is difficult in some cases to know where artist ends and technology begins. Such is especially the case with the Beatles, who, as Tim Riley argues, "are our first recording artists, and they remain our best" (2002, 26). Riley places them on a continuum with Phil Spector as examples of "how studio technique can turn the simple essence of pop into grand, dramatic surges of sound" (2002, 27). Riley says, "A Beatles record is more than just a collection of songs; it's a performance for tape" (2002, 27). It has been noted of the Beatles that for their records the recording studio is played like another instrument. Jerry Zolten places the recording artistry of the Beatles in context of advances in technology in the 1960s in general, saying of artists such as Les Paul and Juan Garcia Esquivel,

> The idea to make studio technology part of music artistry, to consciously apply the medium to the message, flowed from these artists as well as a legion of unnamed rock and roll and rhythm and blues producers. Van Dyke Parks, Brian Wilson, and of course George Martin and the Beatles were caught up in the swell. (Zolten 2009, 36)

Despite what Riley says, what they do there and the ways in which they manipulate sound goes far beyond the mere capturing of a musical performance on tape. And, of course, however we choose to define the artists, they do not create in a vacuum. No art is completely "pure" of motive, and commercial art in particular is caught up in the cash nexus and commodity fetishism that are the hallmarks of capitalism. In fact, not only were the recordings of the Beatles decidedly *not* innocent reproductions of live performances, they were, in fact, by the time of their retirement from the concert stage the farthest thing from it. Zolten says, "[i]n effect, *Revolver* distanced the Beatles impossibly from what they could perform on stage"; on their last concert tour before they retired, "[o]ther than a straight performance of 'Paperback Writer,' their then hit single, the Beatles performed not one song from *Revolver* [their then current album]. They had fully transitioned to the irreproducible" (Zolten 2009, 47). Of course, perhaps the fact that there is no "true Beatles," no "original performance" that the recording could be said to mediate, does not mean that the notion of aura does not have power for collectors of Beatles recordings, it merely relocates the aura of the object in a different register.

Jonathan Sterne notes that to some degree this "philosophy of mediation," by which "copies are seen to be debasements of originals" is a product of discursive practices established in the early marketing of the phonograph. According to Sterne, the sounds a recording captures are not mediated "sounds as they exist in the world" (Sterne 2003, 218). This proposition holds true no matter whether the method of recording those sounds is acoustical, electronic, or digital. Sterne continues,

> The "original" sound embedded in the recording—regardless of whether the process is "continuous"—certainly bears a causal relationship with the reproduction, but *only* because the original is itself an artifact of the process of reproduction. Without the technology of reproduction, the copies do not exist, but then, neither would the originals. (Sterne 2003, 219)

In fact, according to Jacques Attali, the equation of the reproduction with the original is itself the product is a further reduction of music itself to the "commoditization of experience" (Sterne 2003, 242). Attali says, "[m]ake people believe. The entire history of tonal music, like that of classical political economy, is an attempt to make people believe in the consensual representation of the world" (Attali 1985, 46). Attali makes two related points about music and representation that industrialization concretized. First, what Westerners recognize as music is itself an ideological product of bourgeois enlightenment. Attali says "[m]usic, in its all-embracing hope, is simultaneously heard, reasoned, and constructed. It brings power, science, and technology together" (Attali 1985, 61). And, second, that the order imposed on nature by music serves an ideological purpose. By accepting what we understand to be the order of music and harmony as natural, we also buy a conception of social order. Attali says that by gaining the acceptance of such a proposition "the bourgeoisie of Europe finessed one of the most ingenious ideological productions: creating an aesthetic and theoretical base for its necessary order, *making people believe by shaping what they hear*" (Attali 1985, 61 emphasis his). In other words, even in the case of live, in-person musical performance, the question of the unmediated "truth" of the performance is thorny at best, aside even from considerations of the subjectivity of the individual listener entering into the equation. In fact, even when it comes to the plastic arts, the question of mimesis, or likeness between the work of art and its original subject, "likeness" only exists in the art object itself; there is no question unless the work of art itself exists.

Turning again from the irreproducible to the infinitely reproducible, from analog to digital, the introduction of the Beatles to iTunes is in many ways the merger of two great brands. Part of the resistance of Apple Corps to Apple Inc. had to do with the close control the remaining Beatles and the heirs of John Lennon and George Harrison wanted to maintain over all aspects of the brand. Also, Apple Corps and Apple Inc. had to come to a settlement in a long-running copyright infringement lawsuit over Apple Inc.'s use of the Apple name and logo to sell music, since part of the agreement struck between the two when Apple Inc. asked for the use of the name was that they stay out of the music business. According to Apple Inc., "more than 450,000 albums and 2 million individual songs were sold on iTunes worldwide in the first week" (Guglielmo 2014). Because of the resounding success of the Beatles on iTunes, and in honor of the fiftieth anniversary of their debut on the Ed Sullivan Show, iTunes "set up a special channel on its iTunes store devoted to the Beatles with a 14-minute video clip of their Feb. 9, 1964 performance, which kicks off with All My Loving [*sic*]" (Guglielmo 2014).

Despite their years in court, the two Apples seem to have had a symbiotic in addition to a contentious relationship, and Steve Jobs admitted when the iTunes deal was announced that he had had the Beatles in mind even before iTunes was launched. At the press conference, Jobs stated, "we are now realizing a dream that we've had since we launched iTunes ten years ago" (Moren 2010). The gist of much of the press coverage of the announcement was that the Beatles are considered so important in the history of recorded music that offering their catalog provides a legitimating function to iTunes, while iTunes offers the Beatles and Apple Corps new avenues of distribution and perhaps a new audience for the Beatles' music. Which, perhaps, highlights the need to legitimize digital music as an aesthetic object in the first place, the need to establish the "thingness" of the digital sound file. Jonathan Sterne makes the important point that the rise of digital music and "[t]he debate over file-sharing and the MP3's role therein also opens out into a long-standing argument about the nature of music in contemporary culture—as process, as practice, as thing" (Sterne 2012, 185). As this chapter has discussed, recorded music is always already more than once removed from its "performance," or in Walter Benjamin's terms, its "aura" that mechanically reproduced music supposedly loses. However, as Lisa Gitelman writes,

> Records and duplicates always possess quality in the sense of faithfulness; copies can be "true" or "near" copies. Reproductions can be good, bad, or mediocre. They can be better, worse, or equal to one another, but they do not lose the *aura*lessness of their reproductive purpose. (Gitelman 1999, 169).

Gitelman is referring to analog recordings at the turn of the twentieth century, long before the invention of recorded tape and the manipulative possibilities that resulted, let alone the dawn of the digital era with the introduction of Auto-Tune making even further mischief with already contested notions of "aura," "origin," and "authenticity."

In "The Work of Art in the Age of Mechanical Reproduction," Benjamin traced a history of art and aesthetics in which the aura of a unique work of art, its "aura," being a product of its uniqueness, or its authenticity in relation to itself. As Benjamin says, "[t]he presence of the original is the prerequisite to the concept of authenticity." Benjamin further states that aura is destroyed by mechanical reproducibility, arguing that "[t]he whole sphere of authenticity is outside technical—and, of course, not only technical—reproducibility" (Benjamin 1968, 218). Of course, a mechanical reproduction might be closer to or further away from its "aura" in this sense. The more that the performance is manipulated through the use of technology, the more its aura might be said to "whither." The question remains, though, to whither from what exactly? Reproduction in this sense, Attali says

> is the death of the original, the triumph of the copy, and the forgetting of the representational foundation: in mass production, the mold has almost no importance or value in itself; it is no longer anything more than one of the factors

in production, one of the aspects of its usage, and is largely determined by the production technology. (Attali 1985, 89)

The mechanical parts that go into the pressing of a record, though, retain some significance in that they can be used to produce further copies of the work. Furthermore, Benjamin's notion of aura may be more complex than it appears at first. Rajeev Patke quotes a 1936 Adorno letter on Benjamin's notion of "aura," saying,

> Adorno expresses unease about Benjamin's concept of aura, claiming that the idea of aesthetic autonomy is dialectical, and "compounds within itself the magical element with the sign of freedom" (C, p. 128). The issue at stake is the recognition that if "the auratic element of the work of art is in decline," that is "not merely on account of its technical reproducibility," but because "the autonomy of the work of art, and therefore its material form, is not identical with the magical element in it." (Patke 2005)

Despite Adorno's unease, according to Patke, Benjamin's concept of aura serves mainly to highlight the fact that artifice plays a role in the recreation of "nature" in the first place. Patke glosses Benjamin's discussion of the artifice of film as saying that the "process of production is determined by technological intervention" (Patke 2005). Of "The Work of Art in the Age of Mechanical Reproduction," Nick Peim says it is "far from naïve about the distinction between authenticity and reproduction" (Peim 2007, 370). In fact, Patke says, "[a]rtifice thus provides a better illusion of actualization than natural performance, especially with music whose production coincides with its reproductive mode, as in the case of rock music" (Patke 2005). Far from decaying through reproduction, the "decline of aura begins at the origin, with performance becoming art. In the reproduction, the performer and the listener are both alienated from the chronotope of performance" (Patke 2005). In recorded music, in other words, "aura" must refer to something aside from the "natural performance" of the original recorded "event."

The years of wrangling that preceded the Beatles' entry into the realm of online distribution should come as no surprise to anyone who is familiar with their first entry into the digital medium when the program of releasing their digitally mastered catalog on compact disc began in 1987. The Beatles were notably late to embrace CD technology, although perhaps not as late as they were to online distribution. To the problem of locating an original or authentic performance in the analog recordings, digital recording, with its infinite reproducibility adds the further difficulty of control and ownership, as illegal downloading and piracy threaten the existence of the recording industry that made the phenomenon of "the Beatles" possible in the first place. Leaving aside questions of legality raised by this problem, it raises the ontological question of what it is that the commodity that is being stolen actually consists. Is it the potentially infinite number of digital copies that might exist, or the extremely tricky question of what exactly is being digitally copied in the first place? Sterne compares MP3 technology to Heidegger's example of a jug, which "does its holding through the void left by its walls and base." In this comparison,

> The MP3 is defined by the interior space that it creates. The MP3 is a container technology. All communication technologies can be considered container technologies, but the MP3 is a special kind . . . In its conception, the MP3 holds only the sounds that can be heard; it discards the rest and attempts to hide its own excesses. But the MP3 is also a container for recordings—records, CDs, and so forth. It is a container for containers for sound, and it codes the space within itself. (Sterne 2012, 194)

Apple and the Beatles had a problem mostly with their own lack of control over how this space was filled. Sterne notes that "[i]f there is a distinctive 'thingness' to the MP3—and to music in its current digital form—it is that it occupies an ambiguous position that is both inside and outside market economies" (Sterne 2012, 224). The whole thorny notion of "intellectual property" that the struggles over digital content have served to foreground in the past almost twenty years involve notions of the control and power that come with ownership. At the time of the iTunes deal, *The Guardian* said that one sticking point

> on the Beatles' part may have been that the band have always been heavily protective of their music, keen never to devalue the brand by giving away their songs too cheaply: when the disruptive effects of the internet were first felt within the music industry, one common response was to start selling CDs at heavily marked-down prices, but McCartney and co never succumbed to this pressure. (Llewellyn Smith 2010)

In particular, the Beatles did not like the idea of individual tracks from their albums being available for sale, the iTunes model, although they did eventually give in and allow them to be offered. Perhaps money greased the wheels for their eventual entry, though. Calling the deal "more groundbreaking than originally thought," Reuters noted, "iTunes is paying the Beatles' royalties from digital download sales in the United States directly to the band's company, Apple Corps, and is paying the songwriting mechanical royalties directly to Sony/ATV Music Publishing, which controls most of the Beatles' song catalog" (Christman 2011).

With the coming of the Beatles to the digital table came two interesting phenomena that, I propose, are not unrelated to the technologies involved in turning cultural artifacts into salable digital commodities. First, questions arose almost immediately as to the most authentic ways to present the Beatles' recordings. As noted before, EMI Records and George Martin decided to regularize the catalog to the British versions, and to issue the first four Beatles' albums only in monophonic. Previously, the Beatles' labels around the world had packaged Beatles recordings in whatever way they thought could maximize their profits, which meant that albums were broken up and recombined with singles, and altered sonically, including the mono recordings being reprocessed for stereo, to fit the market. In the USA, the aforementioned Dave Dexter—who reportedly did not even like the Beatles' music having passed on releasing it in the U.S. at least three times—took control of their releases, cutting them up and reprocessing the sound, of which Dave Marsh says that it is "indisputable that the group came to hate the idea of Dexter's butchery" (Marsh 2007, 1). However, despite what the remaining Beatles, George Martin, and EMI had thought of Capitol's bastardization of the Beatles' catalog, a move the 1987 digital masters sought to correct, the American audience had grown up with the Capitol recordings, and *they* were considered to be the "authentic" and "original" records for American fans. This situation has been partially "corrected" with the release of two box sets, *The Capitol Albums Vol. 1* in 2004 and *The Capitol Albums Vol. 2* in 2006, which include mono and stereo versions of the US releases from *Meet the Beatles* to *Rubber Soul*. *Volume 2* created controversy of its own, though, when it was discovered that the masters used for the mono versions of *The Beatles VI* and *Rubber Soul* were "fold downs" of the stereo mixes rather than the original mono, a move that was corrected by the red-faced Apple Corps shortly after it was discovered (Womack 2014, 167).

A second, and related to the first phenomenon, the explosion of digital music led to a reassessment of the master tapes, the multitrack recordings that were used to make those masters, and the unreleased recordings remaining stored in vaults. In other words, an incredibly large product of the archeology of the most important popular recordings of the twentieth century had begun. Among other aborted similar projects, EMI had intended to release an album called *Sessions* in 1984 in response to fan interest in unreleased Beatles recordings (Gallucci 2013). However, the release was withdrawn, eventually to be replaced in 1995 and 1996 by the massive *Anthology* series, which promised to be a "warts and all" presentation of the Beatles at work in the studio. Even these recordings were edited and remixed, though, thus presenting only a further simulacrum of the authentic in the Beatles recordings. One review of *Anthology Vol. 2* makes echoes the public reactions at the time of these releases, saying,

> the live tracks sound terrible, and many of the alternate takes are Frankenstein's monsters, assembled from several different uncompleted masters—the same way Alan Douglas worked on Jimi Hendrix's posthumous releases. Personally I would've rather had more really new material, like the entire track of

"12 Bar Original" instead of a three-minute snippet. Once again, fun for the band's millions of fans, but not recommended unless you already have all the original albums. (Wilson n.d.)

As stated earlier, the Beatles' catalog was also remastered in 2009, another massive project undertaken in order to improve the sound using twenty-first century technology and to correct perceived errors that the first digital remasters had made in their representation of the catalog, restoring the original stereo versions of the first four albums and including a boxed set of the mono albums up to the *White Album*. Of course the debates and the criticisms began as soon as the recordings were released. Some believe that mono is the only way to listen to everything from *Please Please Me* to the *White Album*, because the Beatles were present in the studio when the mono mixes were completed, and these mixes are thus more in line with their intentions, although both claims are open to question. Also, despite the claims of Apple that these releases would remain "faithful" to the original analog recordings, some digital noise reduction and limiting was used to make for a more pleasingly contemporary sound, and digital editing was used to correct what were perceived to be errors in the original recordings, most noticeably correcting bad editing in the song "She Loves You." It is these 2009 remasters that form the basis of the iTunes digital releases. Whether the motivations are economic or aesthetic, what it adds up to is the fact that the Beatles project has become one of the archeological excavation and curation of the Beatles recorded legacy, perhaps reinforcing the notion that a legacy exists in the first place.

In the midst of all of this digital and analog plenitude, is this seeming concern with the search for the "original" Beatles recordings the return of the repressed aura that Benjamin was certain mechanical reproduction eventually would render impossible? This reading holds that Benjamin's "narrative of the history of the aura is unsustainable," and that therefore "the essay's account of the death of the aura" is also questionable (Davis 2008). However, perhaps critics of Benjamin's thought such as Jonathan Davis too easily dismiss the power of the framework for understanding the ways in which our view of what art means and what it does changes as a result of mechanical reproduction. Karen Feldman notes Adorno's criticism that "Benjamin's materialist focus on techniques of reproduction and the dissolution of aura neglects the commonalities of auratic and technologically produced art, which represents once again a failure to adequately theorize" (Feldman 2011, 340). The "aura" of a work of art is not the same thing as its "autonomy," to use Adorno's criterion, or, as Feldman says, according to "Benjamin there is no 'appearance of a nonexistent' in the artwork to begin with, whereas for Adorno this is the hope and hallmark of the artwork, where such autonomy promises a life that would not be lived wrongly" (Feldman 2011, 343).

Turning back to the analog recordings of the Beatles, one might assume that the earliest productions would be the closest to the aura of performance. Close they may be, but even these recordings are not raw performances captured on tape, despite the protestations of the technicians charged with capturing those performances on tape. Of the *Please Please Me* album, producer George Martin says, "All we did really was to reproduce the Cavern performance in the relative calm of the studio" (Dowling 1989, 21), while the recording engineer, the late Norman Smith, says,

> I kept the sound relatively "dry." I hated all that echo that everyone was using back then. And I placed the singers' microphones right there with the rest of the band, although singers were usually hidden away in a separate recording booth. I thought that was a bad idea, because you lost the live feel of the session. (Dowling 1989, 22)

However, if the first album tried to capture the live performance as closely as possible, the band soon became aware of the possibilities that the recording studio offered to enhance, and eventually transform, those performances. George Martin overdubbed piano onto several songs on *Please Please Me*. The Beatles also began double tracking lead vocals for a fuller sound very early in their recording career, certainly as early as 1963 and *With the Beatles* (Dowling 50). By the time they recorded *Rubber Soul* and *Revolver*, the Beatles, along with Martin and their recording engineers, were manipulating vocals electronically, using VariSpeed technology to slow down or speed up performances in order to change pitch and alter the keys of the songs, running the tapes backwards, and incorporating sound collages and tape loops into their recordings to the point where it would be extremely difficult if not impossible to locate the aura of the original performance even if access were possible. This history of the Beatles' recordings has been much rehearsed over the years, notably by Tim Riley in *Tell Me Why* and Mark Lewisohn in *The Complete Beatles Recording Sessions*.[4]

Benjamin believed that eventually through the "distraction" caused by endless copies of no longer individual works of art, artistic value in the traditional sense would be destroyed, and that "a generalized equalization of all works of art" would result (Davis 2008). Questions of value would then become entirely subjective. This, for Benjamin, was an antifascist project worth pursuing. However, Jonathan Davis notes that "Benjamin could not foresee that consumerism, rather than communism, would fulfill his antitraditionalist prophecy of the new aesthetic experience." Without collective notions of value are we nothing more than consumers subject to "the impulsive satisfaction of preferences" (2008)? Is the appreciation of the Beatles merely a matter of personal taste? Are the Beatles artists whose body of work withstands the scrutiny of time, or is their work nothing more than a chimera made up of recording techniques and empty discourses of origin? Are the Beatles the group who legitimized mass-produced popular music as an art form, or are they, finally, nothing more than a brand?

Of all of the recent Beatles projects of excavation and curation, the best received have been the mono vinyl remasters, the feeling among the Beatles' fans is that this is the one that "got it right." What they got right, according to fans is the fact that they bypassed the digital mediation that had marred earlier reissues, including the stereo vinyl ones of 2012. The *Chicago Tribune* review of the mono vinyl box makes the case:

> The 2009 stereo and mono CDs used updated technology to create new digital remasters from the original master recordings. The records in the 2012 stereo vinyl box were made from these same remasters, meaning the process essentially moved from analog to digital to analog. Some of the criticism of that boxed set stemmed from the perceived sonic compromises inherent in the digital conversion. (Caro 2014)

The review concludes that this "bypassing of the digital world has a good number of sound aficionados praising the new mono box as the ultimate Beatles audiophile collection" (Caro 2014). *Mojo* quotes one of the mastering supervisors, Steve Berkowitz as saying that the goal on the mono remasters "was simple: make them sound like the original artist and producer intended. We consulted all the original notes. N.A.S.A. should keep such good notes" (Male 2014). According to Haley, "these LPs are marketed to audiophiles and others who believe that digital technology has no place in producing an LP reissue of an analog master tape" (Haley 2015, 151). Two points can be made in relation to this desire, however. The first is that despite the prejudices of some listeners against using the digital remasters for the stereo vinyl, in some cases the digitally remastered vinyl actually sounds better, or as Mark Caro concluded in a sound test at the time, "the new albums fared best" (2012). The second is that the mono vinyl remasters, despite their claims, are not the same thing as the original vinyl records issued from 1962–1968. For one thing, according to Haley, "[t]he master tape for 'Please Please Me' (1962) was too badly damaged to pass through the tape machine continuously for LP cutting, according to the booklet notes. An analog dub was made of the 'Please Please Me' master, and that modern tape was used to cut the LP master" (2015, 153). Also, The new mono vinyl was designed and manufactured with 2015 playback gear in mind, so while the engineers may have followed the original cutting notes and listened to the original vinyl records for reference, the technicians working on the vinyl remasters "experimented with levels and equalization in the context of a modern cutting chain and modern LP manufacturing, which allows for more bass and wider dynamic range" (Haley 2015, 153). If aura has little to do with original performances, though, in this sense perhaps it has something to do with the original recordings themselves. Perhaps, like the CD and iTunes reissues, the vinyl reissues can be said to be "authentic" only to the degree that they borrow their aura from the records the Beatles originally released. Perhaps this question only becomes a question because the original recordings have been reissued in the first place. As Patke says, "[a]uthenticity in a narrow sense might be attached to 'the here and now,' but music permits this to be actualized on a gradient, in repetitions that do not aspire to, or need to aspire to, a first idea of an origin or an original" (Patke 2005). In this sense, that aura might be said to "whither" or "decay" in relation to its distance from the original U.K. mono pressings. In other words, aura is a social construct having mostly to do with the expectations of listening subjects. The records are "supposed" to sound the way they sounded when they were first released. "Dexterization," after all, is only a thing in because the "original" US Capitol recordings exist.

Records, like most other collectables and—in fact like art in general—have value mostly because collectors, curators, and critics agree upon such value in the first place. For collectors of those original recordings such as Jason, perhaps access to the "truth" of the Beatles also does not lie in its proximity to an unmediated performance. Could the reason a first mono pressing of *Please Please Me* is one of the most valuable recorded collectibles on vinyl in addition to being a valuable Beatles collectible be located elsewhere? Farmer perhaps makes a closer approach when he says,

> Not only was most of it recorded in a single day, using less studio time than any other album of the time, but it cost little to manufacture too. Once the takes to be used were decided on, they were edited together to form the production master tape and an EMI engineer (probably Norman Smith, who engineered the recording session) cut the lacquers for both sides of the album. Those first cuts were the ones used for manufacture, signified by the suffixes –IN stamped into the production records being pressed in the Hayes factory. Never again would this be so. Never again would the first cut be considered "good enough" for a Beatles album release. (Farmer 2012, 74)

Farmer raises the not insignificant question in this context, when he asks, "Did some imperfections add to the excitement, or was it just beginner's luck which made *Please Please Me* one of the greatest collections of pop recordings of all time" (Farmer 2012)? Perhaps it is the physical nature of the object itself. Robert Harley quotes mastering engineer Doug Sax as saying,

> The disc has a certain magic. Everything you measure about the disc is worse, except that it has very good phase relationships. I'll tell you what it is. The disc forces the sound into mechanical motion. The speaker is being fed something that's been "predigested" and put into the laws of mechanical motion, which is what musical instruments obey, to start with. (Harley 2007, 39)

Harley suggests that some believe that "the LP in many ways sounds better than the mastertape from which it was cut" a prospect that more or less coincides with Attali's notion that reproductions brings about "the death of the original, the triumph of the copy" (1985, 89).

NOTES

1. Funny Face 26, "Beatles Vinyl Update—Fantastic Grail," YouTube video, 3:01, June 6, 2015, https://www.youtube.com/watch?v=5HhqXr4aAsk&feature=youtu.be.

2. For other details about the various pressings of *Please Please Me*, see the website *The Beatles Collection* (Please Please Me 2011) and Nick Farmer's "*Please Please Me*: Fixing A Hole" in *Record Collector* magazine (Farmer 2012).

3. It should be noted that some of the customer reviews on Amazon.com do not share Kazinn's disdain for the Caruso 2000 recordings. One reviewer even says "Had I read the idiotic review by Allan Kozinn (Sunday, Feb. 20), I probably would not have bought it. What a loss that would have been" (Caruso 2000).

4. Lewisohn, Mark. *The Complete Beatles Recording Sessions: The Official Story of the Abbey Road Years, 1962–1970*, New York: Harmony Books, 1988. This book is still essential reading for anyone with an interest in the recorded legacy of the Beatles.

Chapter Three

Virtual Authenticity

The Reemergence of Vinyl in the Digital Age

Open on a close-up of an album spinning on a turntable. The scene is a makeshift laboratory in Woodbury, GA.[1] At the behest of the governor, Milton is conducting experiments in order to determine whether "walkers" can be either controlled or cured. Andrea assists Milton in Milton's experiment on Mr. Coleman, a dying elderly man, an experiment centered on the question of whether or not the walkers retain anything of their memories and therefore some sliver of their humanity. The record, Jo Stafford's rendition of "It Could Happen to You," is, it turns out, one of the tools Milton uses to facilitate this experiment. At the end of the trial, Milton bends to hear the words the dying man whispers. Andrea asks what he said, and Milton replies that he asked if they could keep the record playing as he dies. The final result of this sequence of experiments is revealed later in the episode to offer no hope, as Mr. Coleman's reanimated corpse can barely be restrained even with the record playing. Of interest to this study is the use of the vinyl record as signifier of a rapidly diminishing humanity. The scratchy surface noise of the stylus in the groove of the record seems both to conceal and reveal certain truths about what it means to be a human being in a situation in which the notion of an "authentic" humanity is very much open to question.

What, after all, do vinyl records and record players, this remnant of the past century, really signify in the digital era? After all, well into the second half of the first century of the information age, disembodied pleasures seem now to be catching up to, if not outpacing, the pleasures of the flesh. Generally speaking, the consumption of music in Western cultures has at least partially transcended the physical media that had been used to carry it for the past 120 or so years. In fact, in the year 2011 digital sales for the first time surpassed the sales of physical media (Perpetua 2012). And, in 2012, "for the first time in history, digital stores became the primary outlet for buying albums, eclipsing mass merchants that had been the leading sales sector for the previous five years" (Gunderson 2013). Make no mistake; digital media are the dominant conveyances of recorded music for the time being. How, then, does one explain the minor resurgence of the lowly vinyl record, a technology that can be traced back to the late nineteenth century, one all but dropped by the major record labels by the early 1990s? Matt Rosoff notes that "[[v]inyl accounts for less than 1 percent of overall music sales, but it's been making a bit of a comeback: sales almost doubled between 2007 and 2008 and grew another 33 percent in 2009, according to Nielsen" (Rosoff 2010). Along with many others, Rosoff observes that the record labels have returned to offering vinyl again, that being one of the few current growth areas in the music industry. Moreover, these journalistic celebrations of the return of vinyl do not even consider the secondary market for vinyl, which accounts for a large part of vinyl collecting and is driven by Internet sites such as eBay and Amazon.com as much as by the physical presence of used record stores all around the industrialized world (Trelease 2013).

The return of vinyl records in the digital age, and the apparent reasons for that return raise interesting questions related to the intersection of technology and aesthetics. In the early days of the introduction of the CD format, audiophiles argued that claims of the superiority of digital sound were specious at best, and for the next twenty or so years, they carried the torch for vinyl, along with alternative and indie musicians and DJs. Such aesthetic claims may be subjective, of course, but the question of the source(s) of such judgments about the superiority of sound can lead to propitious insights about the ways in which culture responds to both familiar and unfamiliar technologies. In this sense, the return to vinyl can be seen as an attempt to recapture an "authentic" aesthetic experience of which the consumption of digital media both via CD and MP3 appears to have robbed us. Concurrently, for many of us, the means we use to listen to music is becoming more and more a mere conduit between "the cloud" and ourselves. Mediated by technology, music itself seems to be more distant and amorphous to listeners, since we no longer even need to have a hard drive on which to collect our music files anymore. What, then, to make of the fact that in the midst of all of this digital plentitude, some listeners are dusting off their old turntables and getting their records out of the attic, sacrificing convenience for the archaic pleasures of the spinning vinyl (or even shellac)? Whether or not one sees this development as nothing more than a hipster fad (as some people do), one cannot help but wonder, why?

The listening subject is also a collecting subject. Journalists and commentators on the vinyl phenomenon make the point that some vinyl consumers do not even own a turntable. In this sense, vinyl records may also represent the aural equivalent of comfort food, an attempt to make sense of a digital world no longer within our grasp. In his book *Vinyl Junkies*, Brett Milano quotes Werner Muensterberger on the psychological need to collect in general:

> Provoked by early, possibly unfavorable conditions or the lack of affection on the part of not-good enough mothering, the child's attempt toward self-preservation quickly turns to some substitute to cling to. Thus, he or she has a need for compensatory objects of one or the other kind . . . To put it another way, such a person requires symbolic substitutes to cope with a world he or she regards as basically unfriendly, even hazardous. (Milano 2003)

Of course, record collecting shares many features with, and therefore resembles, other forms of collecting and might be traced to the same psychological source(s), and collecting is a significant part of the listening experience, but my concern here is with the question of what it is about the vinyl medium that leads people to want to collect records rather than, say, snow globes.

People buy and listen to vinyl records for various reasons, among them the reaction against the corporate recording industry that initially relegated records to the dustbin of history in 1989 (Yochim and Biddinger 2008) . Vinyl collectors derive pleasure from possessing the object, the rituals of maintaining and playing records, and the thrill of the chase of locating hard-to-find titles and pressings, or, as Milano says, "Love *for* the music, love for the artifact, the thrill of the chase: those are three elements that turn a garden variety music lover into a vinyl junkie" (Milano 2003). However, the most commonly cited reason people either never gave up or are now returning to vinyl has to do with "fidelity." Michael Waehner's assessment is typical: "A record in good condition played on a reasonably good sound system will rival or surpass a CD under similar conditions" (Waehner n.d.). A brief visit to an online audiophile forum such as Stevehoffman.tv will tell you that a good part of the chase for the desired item centers on the superior sound reproduction contained therein. The host of the site, mastering engineer Steve Hoffman, works on both CDs and LPs, and he outlines his philosophy as follows, "I want that 'breath of life.' That's what I want. If it sounds like a fake approximation of nothing that's alive—that is not it for me [*sic*]" (Hoffman 2003). While Hoffman admits that he doesn't always accomplish his goal, and that to some degree his goal is an ideal, "lifelike" sound is what he is attempting.

And the term "breath of life" is not accidental; there is an almost religious fervor to the devotion to sound closely related to a format's perceived ability to express "truth." John Davis notes that "[f]or vinylphiles, vinyl's sacred element is strongly associated with a sense of the format's authenticity" (Davis 2007) Davis also takes note of the group of (mostly male) vinyl adherents who, in the face of the digital onslaught, like cloistered monks have kept the flame of analog recordings alive. And the flame they kept until the recent sudden growth spurt of vinyl was and *is* summed up best in the term "authenticity." The definition of the term, however, tends to be circular and self-referential. If one gets too close to the means vinylphiles use to talk about the truth they seek, one notices that it can almost never be discussed without qualification. One always has to account for limitations in the technology used to capture the performance, the various steps taken in the process of manufacture, and the equipment used to reproduce the sound. Hoffman himself says, "I have actually figured out how to make the LP cut and the master tape sound EXACTLY the same" (Hoffman 2003). The "breath of life" in this case, however, is not fidelity to the original performance, if such a thing ever existed, but rather fidelity to the *tape*.

How exactly does such authenticity of sound reproduction relate to questions of authenticity of the sounds reproduced and the objects on which said reproductions are inscribed? It is not so much that the question of authenticity is raised in relation to music's emergence in the virtual realm to begin with, but rather the desire among some listeners to reclaim a notion of authenticity that they feel is lost as sound becomes digital. Yochim and Biddinger relate this desire to what they call the "trope of death" embedded in analog recording discourses from the very beginning, pace John Durham Peters and Jonathan Sterne. According to Yochim and Biddinger, "the moment that CDs challenged vinyl's dominance, the death metaphor re-emphasized the connection between vinyl and humanity. At this moment, people began regarding vinyl as precious and revering it" (2008, 185). Just to be clear, authenticity per se resides in neither original performance nor recorded copy thereof, which is not to deny that the question of authenticity is legitimately raised. Sterne says, "[s]ound reproductions that are acknowledged as wholly artificial by performers or the audience (or both) can still come to have a sense of authenticity. But this notion of authenticity refers more to an intensity or consistency. It is a claim about affect and effect rather than a claim about degrees of truth or presence in a reproduced sound" (2003, 141). Neither is the desire for authenticity to be taken for granted, since it is every bit as constructed by and implicated in cultural relations as are the objects by which it seeks to authorize itself.

In fact, at various points in the history of recorded music, the desire for what listeners thought of as authenticity has been responsible for moments of renewal and the creation of new genres of music. Hugh Barker and Yuval Taylor begin their study, *Faking It: The Quest for Authenticity in Music*, at the intersection of at least two such historical moments, with Nirvana and performing Leadbelly's "Where Did You Sleep Last Night" on MTV's *Unplugged* show in 1993 (Barker and Taylor 2007, 2). Barker and Taylor discuss this moment in terms not only of the musician's quest for authenticity of expression but also the audience's expectation of authentic representation, and they locate these impulses in the late 1980s and early 1990s with the emergence of grunge music and also in the 1930s, with folklorists John and Alan Lomax and their quest for what they thought of as authentic African-American music, which Barker and Taylor claim they associated mostly with "the most primitive, elemental, and backward-looking African-American musicians they could find" (2007, 12). It is important to note both that at the time of and despite the Lomax's promotion of Leadbelly, "African-Americans were creating much of the most sophisticated (i.e., complex, non-primitive) music in the country, from blues to rhythm-and-blues to jazz," and also that the music that John and Alan Lomax collected was enormously influential on both later developments in music, such as the folk boom in the 1950s and 1960s, and on attitudes about music as well. In this sense, Barker and Taylor suggest that Kurt Cobain's choice of a "closing statement," not only for his appearance on MTV's *Unplugged* series, but also for his career and his life, was not accidental. Cobain believed that to make an authentic musical statement, an authentic music was a necessity. Barker and Taylor suggest that Cobain and grunge painted themselves into a corner in their quest for authenticity, to the ultimate destruction of both. Rock musicians, like the folk musicians before them, believe that the authenticity they seek can be found in music such as the music Leadbelly made. Of their recourse to the music of Leadbelly, Barker and Taylor say,

> Leadbelly's appeal, not to mention his fortune and fame, both when he was alive and after his death, derived from a racist view that the most authentic black culture was the most primitive. Over time, this view has lost most of its racial tinge: now it is commonly accepted by rock fans the world over that the most authentic music is the most savage and raw. (2007, 38)

"Even the most transparently manufactured pop musicians," Barker and Taylor conclude, "look for fig leaves of authenticity" (Barker and Taylor 2007).

Barker and Taylor ultimately make the point that the kind of musical authenticity sought by fans and folklorists alike cannot be attained, but that fact does not diminish the desire for authenticity. They say, "one can't always put authenticity to one side. Tastes formed on the base of authenticity change only slowly, and many old prejudices survive" (Barker and Taylor 2007). They note that the relationship in the listener to moral and aesthetic questions of what is "real" are motivated largely by the connection between music and identity that forms when we are young. Even absent the justification of the authenticity of the musical performance itself, however, listeners still crave authenticity. Jonathan Sterne says that "[s]ound reproductions that are acknowledged as wholly artificial by performers or the audience (or both) can still come to have a sense of authenticity. But *this* notion of authenticity refers more to an intensity or consistency of the listening experience" (2003,

241), which may just bring the discussion back around to the reemergence of the analog turntable in the early twenty-first century. Travis Elborough makes the point that the LP album "has only survived this long because it met the needs and fed the desires of its times" (2009, 398). Elborough connects the reemergence of vinyl specifically to the changing ways in which it is conceived as a marketable item. The record store has also returned, but in a new and different register: the contemporary record store offers "a journey into . . . well, some serious space. And customers are encouraged to treat it as *a* space" (Elborough 2009, 398). Bartmanski and Woodward amplify this point, saying that such stores "represent the capacity of vinyl to reach into expanded and reinvigorated markets and the cultural affordances of vinyl to become entangled with other cultural consumption goods, though they are also spaces where boundary maintenance around matters of perceived authenticity and mainstreaming come to be exercised" (2015, 156). Elborough notes the analog record's connection in common perception of what it means to be human, to the point that a gold-plated record and the means with which to play it were sent into space on the Voyager space probes to greet alien space travelers (2009, 394). Could it be said that the return of vinyl and the record store both are attempts to bring music back to Earth in the information age that would treat it as disembodied data floating through cyberspace? Which raises the question, do we seek for our "true" and "authentic" selves in the junk pile of outmoded technologies?

Miles Orville makes this point, saying, "[o]wning things, even mass-produced things, has long been recognized as a way to rebuild or change the self, to arrive at a more 'authentic' sense of selfhood" (2014, xviii). Orville contrasts this search for authenticity with the "culture of imitation" that prevailed in the nineteenth century, in which, according to Orville, "the new force of mechanization was identified with the spirit of national progress" (2014, 33). Orville describes what could be the impulse behind the analog record when he says,

> [o]ne dominant mode in poular culture in the late nineteenth century was thus the tendency to enclose reality in managable forms, to contain it within a theatrical space, an enclosed exposition or recreational space, or within the space of a picture frame. If the world beyond the frame was beyond control, the world inside of it could at least offer the illusion of mastery and comprehension. And on a more elementary aesthetic level, the replica, with its pleasure of matching the real thing and facsimile, simply fascinated the age. (2014, 35–36)

At the turn of the twentieth century, though, in response to that culture of imitation and "continuous with the impulse of the younger generation of realists in the 1890s," there arose a "new culture of authenticity," which "derived its form more specifically as a response to the vast consumer culture that was implacably taking shape" (Orville 2014, 180). Orville points to ways that modernism in art, design, and literature embraced "reality" as a means of rediscovery, noting in his discussion of William Carlos Williams, "[m]aking the artwork real and making it new meant overhauling the language of description and breaking open the closed forms of literature in a way that was consonant with the new facts of modern life, but it also meant restoring what was thought to have been taken away—contact with reality" (2014, 280).

Orville says that for the modernists, the real world required a new type of seeing, one influenced by the advances in technology of the late nineteenth and early twentieth centuries, and this new way of seeing "was in fact an openness to the formal implications of the very things that were changing 'how everybody was doing everything'—an openness to popular culture, to science and technology, above all an openness to the forms of visual representation that were evident in photography, advertisements, and the cinema" (2014, 282). The focus for modernism shifted, however, from the authenticity of the thing represented to the authenticity of the form that represented it. This shift, and the attendant representational confusion it attempted to resolve ultimately resolved little, but it did make its way discursively into the wider popular culture in demands for further "realism" in cultural products such as records and films, in turn rendering "authenticity" as a category of aesthetic judgment very problematic indeed. In the case of recorded music, the question remains, "authentic to what?" The amorphous nature of the recorded performance, its ontological slipperiness, leaves little on which to hang a claim, to say nothing of the various formats by which it has made its way into the market.

However, the search for authenticity seems to be extremely well-suited to the technologies involved in making recorded music, or it could be that recorded music and its attendant technologies have contributed to modern and post-modern anxieties about authenticity, "the tension between imitation and authenticity" that conditions "our aesthetic and material culture" (Orville 2014, 326). Orville notes that given such tension, in post-modern culture "it is not surprising to find a strong undercurrent of fascination with junk" (Orville 2014, 326). Orville highlights the creative possibilities of junk, the act of "starting with something old; it is the object found and rescued, reclaimed, reworked, reintegrated, the thing with a history, the mass-produced object become individualized, the object to be collected" (2014, 326). Orville could be talking about vinyl record collectors rather than the photographs of Walker Evans when he says that Evans,

> took the mass-produced thing, identical with its thousandth mate off the assembly line, and restored its individuality, the product of its implied history. He took the reproduction that, as Benjamin would have it, was lacking in the

charisma of an original "aura," and endowed it with originality. He took the thing that, as junk, no longer speaks to us in its original voice, and gave it a new voice in the context of dissident values. (2014, 328)

Vinyl listener Erik Thompson illustrates this point, noting that in a world of digital immediacy and consumption, "[c]omplete musical artistic statements still matter to a select few of us, and a vinyl record provides a tangible, illustrated entry point into a musical world" (Thompson 2015). Thompson connects the technology to the idea of humanity, noting that "[i]n this On Demand day and age, where everything can be controlled by remote, Amazon orders (including vinyl orders, ironically) arrive in an hour, and cars can drive/park themselves, it's both rewarding and humanizing to play a small part in any creative experience, especially one that matters to me as much as music does" (Thompson 2015). There seems to be here a self-conscious attempt to reclaim one's humanity by adopting the older technology. Or, perhaps a better formulation might be to call it a self-conscious attempt to create one's "reclaimed" humanity through the use of this discredited format. This is not, at least for Thompson, an exercise in nostalgia, nor is it an attempt to seem hip. Instead, he says, "I truly believe music sounds richer and fuller on wax, and after a day spent immersing myself in the polished purity of digital sounds, it's gratifying to hear songs I love in an analog format while holding the sleeves and jackets of the albums themselves" (Thompson 2015). Thompson says, "Music becomes a far more personal experience for me on my home stereo with my turntable spinning, and that sense of intimacy brings me closer to the heart and emotion of the songs. For that moment, the outside world fades away and the music is everything. The only thing. The way it should be" (Thompson 2015).

The way it should be usually means as close to the original recording as possible. Processing or manipulating the recorded sound in any way tampers with its authenticity according to collectors such as the Korean sound engineer who blogs as Analog Jazz. For this blogger the best way to listen to recordings made before the 1980s is "the 1st press vinyl which used the original master" (Why? 2013). Analog Jazz mentions the degradation of magnetic tape as one of the reasons to avoid newer remasters of old analog recordings, but he also notes that the process of manufacturing recordings can interfere with the fidelity, and therefore authenticity of the sound. As he says,

> The manufacturing process of a vinyl is a mechanical process. The stamper has an inverse groove image of the record and it is pressed against hot vinyl. The problem is that the stamper gets worn out after several hundred stamps, and was usually replaced after 1000 or so. This is a small but important reason why 1st press is best in terms of sound quality.(Why? 2013)

Scott Miller examines the claims of the character Rob and his cohorts in the musical adaptation of Nick Hornby's *High Fidelity*, claims that are not far from the claims made by actual vinyl aficionados, saying that for them, "*fidelity* means more than faithfulness to sound quality; it means faithfulness to music as art, as spiritual experience, as life itself. They cling to the technology of LPs as many audiophiles still do today, believing that the sound is warmer and realer than the digital sound of CDs" (2011, 234). Travis Elborough equates the return to vinyl in the face of more "advanced" technologies to "an act of historical reenactment, like putting on a stovepipe hat and pretending to be a Parliamentarian pike carrier at weekends" (2009, 13). He also relates this return to the relatively recent phenomenon of "classic" artists playing complete albums in concert,[2] saying, "Artists and audiences at these concerts appear locked into mutual bouts of nostalgia for the days when everyone felt a greater obligation to listen to every track at home—an obligation that iTunes and chums have helped to dash" (Elborough 2009, 14). Whether this recent concert phenomenon is related directly to technological developments in digital music technology might be open to debate, but Elborough makes the not insignificant point that streaming and digital downloads have limited musicians' ability to make a living off of recordings of their music to a large degree, saying, "it has become a far more reliable and financially rewarding stream of income for many artists" (2009, 14).

In a sense, the quest for authenticity that brings listeners in the opening decades of the twenty-first century to return to the vinyl analog record must be seen as at least partially driven by nostalgia. If it is nostalgia, though, it is of a particular type and in response to a specific set of circumstances. Albert Borgmann refers to "the dominant way in which we in the modern era have been taking up with the world," a way he calls "the device paradigm" (2009, 3). Borgmann illustrates this "determining pattern of our lives" using the example of a "stereo set,"

> consisting of a turntable, an amplifier, and speakers, [it] is a technological device. Its reason for being is well understood. It is to provide music. But this simple understanding conceals the way in which music is procured by a device. After all, a group of friends who gather with their instruments to delight me on my birthday provide music too. A stereo set, however, secures music not just on a festive day, but at any time, not just competent flute and violin music, but music produced by instruments of any kind or any number and at whatever level of quality. (2009, 3–4)

The connections between the devices used to reproduce the music and the music reproduced are invisible, or at least not immediately apparent, according to Borgmann. Borgmann calls the distance that the device paradigm puts between human beings and the "focal things and practices" that give our lives meaning and authenticity, that "center and illuminate our lives" (2009,

4). He continues, "Music certainly has that power if it is alive as a regular and skillful engagement of body and mind and if it graces us in a full and final way" (2009, 5). Borgmann's study is an argument for the return to those focal things and practices from which modern technology and the device paradigm have distanced us. In his discussion of Heidegger, Borgmann raises the question "how are we to recover orientation in the oblivious and distracted era of technology when the great embodiments of meaning, the works of art, have lost their focus and power" (2009, 203). Heidegger suggests that "we must uncover the simplicity of things," and, furthermore, that "if we recognize the central vacuity of advanced technology, that emptiness can become the opening for focal things" (Borgmann 2009, 204).

The search for "happiness" in the face of the indifference of technological society is Borgmann's quest, and the return to focal things is his means of getting there. Gordon G. Brittan Jr. makes this quest explicit when he says, "Borgmann is attracted to the idea of excellence. . . . But he realizes that it involves a set of ideals produced by a culture that was not 'device-ridden' in the same way as ours and thus might be taken as an alien and unfair standard by which to judge contemporary technological life" (2000, 78). Because excellence is not open to him, Borgmann must resort to happiness as the standard by which he judges technological culture and the device paradigm wanting. For our purposes might there not be an intermediate standard of excellence or of happiness between focal things and devices? Brittan argues, however, that happiness is a slippery standard at best with which to form such a judgment, noting, "it might be argued that whatever dissatisfaction is felt with the technological culture is a function of its having been achieved at the present moment and, never content with what is at hand (in fact, rather bored with what is at hand), we look to the past (more real than the future) for a happiness that, not now within our grasp, seems for that very reason more perfect" (2000, 82). Brittan's main line of counter-argument against Borgmann seems in the final analysis to hinge on the fact that Borgmann's account of technological society and focal things allows for little middle ground between the two poles. In the end, he suggests, "we do not have to renounce or withdraw from our technological culture if at the same time we want to lessen our dependence on it, to become more dependent on ourselves and on each other. What we do have to do is keep alive, individually and socially, a range of basic skills, if not also a set of 'focal practices'" (Brittan 2000, 84). Human beings in advanced technological societies seek and find spaces within those technologies in which to express their freedom and independence, in other words. The skills and competencies required to indulge a taste for vinyl analog audio might not be quite up to the same standard as learning to play the violin in Borgmann's terms, but they do provide more of a "human touch" than listening through earbuds to Spotify on a smart phone app.

Fidelity can mean many things in the case of recorded music; it can mean the faithfulness of the recording to the initial performance, something akin to Walter Benjamin's use of the term "aura" in "The Work of Art in the Age of Mechanical Reproduction;" it can mean the fidelity of a particular reproduction (or pressing) to that performance; it can also mean fidelity to the truth of the listener's lived experience. Brett Milano quotes a vinylphile who contrasts the true sound of vinyl with the "soulless" sound of CDs and who takes the concept of fidelity to mimetic extremes: "[m]usic is all about the groove, and you don't find grooves on a CD" (Milano 2003). In an audiophile forum discussion of the best pressing of Elton John's *Goodbye Yellow Brick Road* LP that includes debate over the "presence," the "life," and the "soundstage" of the music contained in the grooves of the variety of pressings available, one member quotes a review of an audiophile pressing in which the reviewer notes that the sound is so true that he can hear Elton John's feet on the pedals of the piano! While there isn't space in this chapter to refute such claims of fidelity to the source, recording engineers note the need for the use of compression of the dynamic range in order to overcome the limitations of the vinyl medium, so any way you slice it, there can be no such thing as truth to the sound of the performance, even if there was such an unmediated thing (Dankosky 2012). Also, given the fact that many recorded performances are "comped," a process whereby a single composite "performance" is stitched together from many recorded takes. For example, "A singer records the song through a handful of times in the studio, either from start to finish or isolating particularly tricky spots. Starting with between 4–10 takes is typical—too many passes can drain the artist's energy and confidence and also bog down the editing process later" (Neal 2015). After all of the takes are recorded, "The recording engineer picks out the best take for each bit of the song and edits all the pieces together, usually in a digital audio workstation (DAW) like Pro Tools (Neal 2015). The process of comping did not begin with digital recording technologies, however, as Travis Elborough notes in his discussion of Glenn Gould, who retired from live performance explicitly because it could not capture the perfection of the performance in the same way a composite recording could. According to Elborough, "[t]he quality of a recording, not its authenticity, was the only criteria that mattered to [Gould]" (2009, 60).

Furthermore, because of the wear and tear on all of the various parts involved in the complex process of manufacturing a record, replacement parts create different recorded products, called "pressings" in the recording industry. Generally speaking, the more "authentic" pressing of a given album is the one pressed first and in the artist's native country, because of the proximity of the original master tapes. But even finding the "true" pressing can be a daunting process, given the fact that so many variables are at work, a subject to which chapter 4 will return in more detail. One of the best selling rarities in the Beatles catalog is the first Parlophone pressing of *Please Please Me*. In an article produced for *Record Collector* magazine, Nick Farmer notes that because of the variations in the matrixes, stampers, and mothers, not to mention variations in the printing of labels and sleeves, it is difficult to decide even what constitutes a first pressing (Farmer, 2012, 76). Suffice it to say for now that even not all first pressings were created equal, and that while collectors spend time and money chasing for such illusive treasures, like Schrödinger's cat, what they wind up holding in their hands both is and is not an item unique unto itself and therefore authentic.

If the preceding types of authenticity slip away from our grasp the closer we examine them, perhaps vinyl records express a greater fidelity, the authenticity of the experience of being a vinyl listener. In fact, our perception of sound is to a great degree dependent on our enculturation. Acoustic researcher Sean Olive makes this point, saying

> Well, there's a number of factors involved in our perception of sound quality, and a lot of them have nothing to do with the sound itself. So in the research that we do, we've looked at the—and we call these things nuisance variables or biases. So one of the ones that we deal with is psychological nuisance variables, which have to do with your knowledge or expectation of what you're hearing. Thomas Edison knew this a hundred years ago. He said people will hear what you tell them to hear [*sic*]. (Dankosky 2012)

Vinylphilia in this regard defines itself to some degree against the perceived dehumanizing aspect of digital technology. The resurgence of vinyl shares its structure of feeling with other popular current entertainments, post-apocalyptic narratives such as *Revolution* and *The Walking Dead*, both of which imagine a world of human agency in which digital technology no longer holds sway. In fact, metaphorically speaking, who are the "walkers" in *The Walking Dead* if not the hordes of the plugged in, their eyes glued to their smart phones and earbuds plugged into the cloud, eating the last vestiges of humanity alive?

"Authenticity" might perhaps be better defined as an episteme, something akin to Raymond Williams' "structure of feeling:"

> We are talking about characteristic elements of impulse, restraint, and tone; specifically affective elements of consciousness and relationships: not feeling against thought, but thought as felt and feeling as thought: practical consciousness of a present kind, in a living and interrelating continuity. We are then defining these elements as a "structure": as a set, with specific internal relations, at once interlocking and in tension. Yet we are also defining a social experience which is still in process, often indeed not yet recognized as social but taken to be private, idiosyncratic, and even isolating, but which in analysis... has its emergent, connecting, and dominant characteristics." (Williams 1977, 132)

Williams distinguishes "emergent" cultural meanings and values from what he calls "residual" and "archaic" ones. If the current digital revolution might be seen as "emergent," then perhaps authenticity as a value grounded in analog technologies could be seen as "residual," meaning it "has been effectively formed in the past, but it is still active in the cultural process, not only and often not at all as an element of the past, but as an effective element of the present" in Williams's terms (Williams 1977, 122). For vinyl listeners, it is the very imperfections of the technology that help create the rituals that constitute the authentic, listening subject, which they perhaps connect with a prelapsarian, more innocent time. One such vinyl listener, asked about his preference for this study, responded, "I think my vinyl records 'fandom' comes from the womb. There was always a console stereo at home and I was obsessed with it and those black discs. One of my uncles had a record collection that I loved immensely and it was him, without even noticing, that made me love vinyl records the way I do" (A. Martín Gómez Acevedo, personal communication, January 2, 2015). Another listener cites a "sort of 'organic' sense to vinyl playback when everything is optimal" (B. Wood, personal communication, January 2, 2015). This same listener says of how he acquired a taste for vinyl, "my father acquired a set of RCA Red Seal LPs with exemplary recording quality and great performances." The search for the most authentic pressing to be played and maintained in the optimal way is the quest of the vinyl collector, which is both a social and a private phenomenon, and, while this has always been the case, the invention of digital technologies has inflected this quest in significant ways, not least of which is providing the ground for further dissemination of formerly obsolete technologies such as turntables and records. This is a topic that the next four chapters will discuss in greater detail.

The somewhat hirsute resurgence of vinyl also roughly coincides with what *Entertainment Weekly* terms the "folk-rock revival" embodied in the popularity of such bands as Mumford and Sons and the Lumineers. Mumford and Sons in fact registered "the highest first-week sales for a rock band since 2008," going so far as to compare the phenomenon to "the popularity of DIY craft hub Etsy to the prevalence of farm-to-table restaurants" (Maerz 2013, 40–42). It should be noted that all of the bands discussed in this article make it a point to release their records on vinyl as almost the preferred format. The article even has a member of Old Crow Medicine Show bragging that Doc Watson called their music "authentic" (Maerz 2013, 41). Even in this case, however, authenticity becomes illusory, as Melissa Maerz notes "nobody in this scene is going to swear off technology. Of Monsters and Men regularly use synthesizers; Bon Iver occasionally tweaks his vocals with Auto-Tune" (2013, 43).

One venture in the digital realm that explicitly connects itself to both analog technology and authenticity is the online record subscription service

Feedbands, which describes what it does as, "a vinyl record subscription service and streaming service that works exclusively with independent musicians. We release a new vinyl record each month from an artist who has never had their album pressed to vinyl. When we release an artist on vinyl, the artists get cash and records and keep all the rights to their music (Feedbands, 2015). In their announcement, Feedbands even makes a direct connection between the authenticity of the music and analog technology to the authenticity of food, announcing the September 2016 launch of "Farm to Feed Bands," located in Ashville, NC, "a farm where bands can sleep and eat free of charge and even play a show if they want" (Feedbands 2015). The Feedbands website describes what they do in a way that highlights the wholesomeness and honesty that they would like to connect to their main business of selling vinyl records, saying,

> All food grown on the Feedbands Farm is grown from organic seed. No synthetic pesticides or fertilizers are used. We are dedicated to growing the cleanest, most nutritious, most delicious food humanly possible. In our first year experimenting as farmers we have successfully grown tomatoes, broccoli, kale, chard, amaranth, corn, zucchini, butternut squash, acorn squash, blueberries, hardy kiwis, strawberries, okra, beets, carrots, radishes, and more and now we're ready to open it up to bands. (Feedbands 2015)

Raymond Williams could be describing this project when he says, "the idea of rural community is predominantly residual, but is in some limited respects alternative or oppositional to urban industrial capitalism, though for the most part it is incorporated, as idealization or fantasy, or as an exotic—residential or escape—leisure function of the dominant order itself" (Williams 1977, 122). Feedbands offers the farm as a place to eat, sleep, and play for independent musicians and bands, and Feedbands considers what it does to be in direct opposition to the practices and offerings of mainstream musical culture. Despite the hands-on, organic, and authentic tone of Feedbands' marketing, however, it goes almost without saying that the subscription also includes a download code for digital versions of the records it offers for sale.

If it does not create, the folk-rock revival reinforces the creation of a virtual authenticity that it shares with the vinyl aesthetic that might be better understood as an aural equivalent to Jacques Lacan's definition of *objet petit a*, the unobtainable object of desire through the search for which we constitute our subjectivity. Fidelity must always be the desired object, but misrecognized and unattainable. Sean Homer defines the gap thus,

> It is this gap that inaugurates the movement of desire and the advent of the *objet petit a*. Through fantasy, the subject attempts to sustain the illusion of unity with the Other and ignore his or her own division. Although the desire of the Other always exceeds or escapes the subject, there nevertheless remains

something that the subject can recover and thus sustains him or herself. This something is the *objet a*. (2005, 87)

Consequently, "The objet a is both the void, the gap, and whatever object momentarily comes to fill that gap in our symbolic reality" (Homer 2005, 88) . Expensive vinyl reissues, some in elaborate packaging, beckon discerning listeners with their historical significance, offer themselves as means to fill that gap, thus constituting the listening subject as discerning in the first place. Your experience is "authentic," because you have the record to prove that it is. Brett Milano spotlights collector Clark Johnson in his book *Vinyl Junkies*, a man who's "Boston house easily has a half a million dollars' worth of music in it." Milano notes that Johnson "still keeps the first record he ever owned, a disc of his own voice made in one of those old amusement-park booths" (Milano 2003). Dropping the stylus onto the record, through the pops, the clicks, and the surface noise, we strain to hear the "authentic" sound of our own voices.

NOTES

1. "When the Dead Come Knocking." *The Walking Dead*. AMC Network, November 25, 2012.
2. Having attended a few such concerts, most recently Steely Dan performing *The Royal Scam* completely as part of a multi-night, multi-album stand at the Beacon Theatre in New York City, the author can attest to both the popularity and the power of such performances for the classic rock audience.

Chapter Four

Criminal Records

Record Collecting as Counter-Discourse

The animated science fiction film *WALL-E* is about what is left of the planet Earth in the year 2805. The human beings have all escaped the waste-covered planet into outer space, leaving behind trash-compacting robots tasked with cleaning the refuse of their rampant consumerism and disregard for their environment (Stanton 2008). As the film begins, WALL-E appears to be the only robot of his kind left, and, with his cockroach friend for company, he spends his days on his impossible task and his nights in a makeshift home he has created for himself that he populates with objects that he collects from the dusty piles of trash through which he sifts. In the seven hundred years he has been fulfilling his duties, WALL-E has somehow developed sentience, and he whiles away his loneliness by watching a VHS tape of the musical *Hello Dolly* that he has discovered, particularly the song "It Only Takes a Moment." WALL-E learns about companionship by watching the characters in the film holding hands while singing the song, and he uses the song to teach the robot EVE the same lesson after she is dropped on Earth by the humans in space in a search for signs of viability. WALL-E uses the refuse that the humans leave behind to create a sort of counter-discourse to the discourses of crass consumption and utilitarian necessity industrial capitalism. That the robot has a differing system of value is signaled not only by the use to which he puts refuse of consumption represented by the videotape but also by the objects that catch his eye. For instance, at one point he finds a jewel case containing a diamond ring, and he throws away the jewel and keeps the case. In this chapter, I am going to argue that record collectors create a similar counter-discourse by valuing the records they collect in a way that differs from how they are told to by the discourses of the record industry that produces the commodities they collect.

In *A Grammar of Motives*, Kenneth Burke remarks that, "Men seek for vocabularies that are reflections of reality. To this end, they must develop vocabularies that are selections of reality. And any selection of reality must, in certain circumstances, function as a deflection of reality" (Burke 1945, 59). Chapter 1 asserted that, apart from any actual correspondence between the sound recorded and the recording of the sound, the notion of "fidelity" in fact comprises a discourse whose language and practices help to constitute the listening subject. This chapter examines the language used in the discourse of vinyl collecting, specifically the ways in which the language that vinyl listeners use functions to interpellate them as collectors, the realities it reflects and the realities it deflects. In fact, the first three chapters of this book have examined the reasons why an individual might choose vinyl over other formats for listening and collecting. This chapter looks at how and why a listener might justify his or her choice, and it also looks at the ways in which vinyl collectors talk to each other and how the language used to talk about vinyl records contributes to the ways it is thought about.

Contemplating his book collection, and talking about collecting in general, Walter Benjamin said, "there is in the life of a collector a dialectical tension between the poles of order and disorder" (Benjamin 1968, 60). Making sense of this tension requires specific ways of thinking and talking about it. Kenneth Burke calls these ways of thinking and talking "terministic screens," which determine the ways we might look at the same object as if through different filters. The language of vinyl collecting provides one example of a terministic screen "directing the attention and shaping the range of observations implicit in the given terminology" (Burke 1966, 50). These screens show what can be "seen" and "known," but at the same time they also block other things from sight and from knowledge. The language of collecting in general usually seems to center on questions of order and of value. The term "value" takes on an interesting hue in the case of talking about something as lacking in value to most of the general public as vinyl is. Furthermore, there is a duality related to the dual nature of the record itself, as both performance and as object that inflects notions of value. About the reports of the subjects of his study of vinyl record collectors, Roy Shuker says, "use value and exchange value [are] commonly held in tension. Those who [claim] a love of music as central to their collecting [are] also proud of items they had paid high prices for, or [are] very valuable" (Shuker 2010, 53).

Record Collector magazine, the British version, perhaps, of *Goldmine* in the USA, includes a monthly feature titled "The Collector," an interview of a prominent individual talking about his or her record collection. These interviews commonly conclude with a question about what the subject will do with his or her collection after he or she dies, suggesting perhaps the idea that even having a record collection rests on the question of what will become of it. Record collecting can be said to be an activity that comprises at least partially the collective traces left behind by consumer culture. That the collection eventually leaves traces itself complicates the ideas of disposal and preservation considerably. Stanley Cavell quotes Benjamin as saying "living means leaving traces" (Cavell 2013, 106), and later, in Cavell's discussion of collections of items of natural history as opposed to art and artifacts, he notes,

> In both arenas of display death is invoked, even death as in life, but in collections of art and artifacts it is my death that is in question as I enter the stopped time of objects (Pomian remarks that their display is as on an alter), whereas the skeletons and parts of natural history speak of the death and perpetuation of species, of their coexistence and succession, measured within the earth's time. (Cavell 2013, 121)

And, what of these record collections? Will the traces left behind give meaning to the lives of the collectors who amassed them? The collectors in *Record Collector* offer at least two alternatives. Neil Campbell answers, "My son has just asked if he can have them, so he may take up the mantle" (Campbell 2014), while Ashley Beedle responds, somewhat facetiously, "I'll have them thrown into my grave with me, scattered on top of my coffin like a Viking burial. They're coming with me, over to the other side . . . Sorry kids, but that's your inheritance down the Swanee" (Beedle 2014). Interestingly, both responses hinge on the idea of inheritance, in the need to extend oneself, through one's collection, into the future, because as objects, records belong to the historical record of the past, but at the same time, records contain preserved performances of the past, so, as objects on which to project one's identity, they represent a threefold past, the past of the collector, the past of the object collected, and the past of the performance contained by the object collected.

Roy Shuker cites a study on collecting by Pearce, which identifies three distinct modes of collecting in general that Shuker believes apply to record collecting. Shuker says, "each of these collecting modes are represented among record collectors, broadly corresponding to varying emphases: on recordings as part of identity formation and life history (souvenir collecting); accumulation and completism (fetishistic collecting); and discrimination and connoisseurship (systematic collecting)" (Shuker 2010, 7). While Shuker rightfully makes the point, often made by other theorists of both record collecting and collecting in general, that it is difficult if not impossible to assign one impulse or motivation to record collecting, of the difference between music fans and collectors he notes, "[w]hile fans will collect records, record collectors are more often characterized by what can be termed 'secondary involvement' in music, activities beyond 'simply' listening to the music: the seeking out of rare releases, such as the picture discs and bootlegs; the reading of fanzines in addition to commercial music magazines; concert going; and an interest in record labels and producers as well as performers." (Shuker 2010). Dominick Bartmanski and Ian Woodward say, "certain artistic commodities retain polysemy that enables different carrier groups to subvert the very logic of master narratives that seem to govern the production and reception of culture" (2015, 101). Berk Vaher defines what he calls "the worldwide vinyl community," calling it "a movement in praise of the vinyl record as an iconic material vehicle of memory in twentieth-century culture, flourishing in the face of the digital recording revolution of the last decades of the twentieth century and expanding in the download 'noughties' of the new millennium" (Vaher 2008).

The other question often asked in the profiles of record collectors in *Record Collector*, not unrelated to the one about the eventual disposition of the collection, is "What is your rarest/most valuable item?" Both questions assume a certain significance in relation to the specialist areas of each of the collectors profiled in that they elicit responses illuminative of the subjective value the interview subjects place on their collections. Collecting vinyl records is in itself a specialization perhaps difficult to justify to the non-collector, let alone some of the specialized forms of collecting profiled in the magazine, such as DJ Jim Backhouse, who says that he is "drawn to sounds that strive beyond the limits of genre or taste," an example of which is "Monoton's *Blau*," which Backhouse describes as "a wonderful set of minimalist dub electronics from 1980 that's uncannily ahead of its time" (Backhouse 2014). The question of value becomes a slippery one under the weight of the divide between subjectivity and objectivity, especially in a case in which the value of the object depends to the collector on the fact that it is *not* valued by other collectors. Some collectors share the opinion of Erik Stein, who, when asked about the worth of his collection, answers, "That side of collecting doesn't interest me much. It's more about the obscurity and the threads linking different bands, scenes, and labels" (Stein 2014). In fact, many of the collectors profiled will answer that question in a similar fashion.

Vaher calls these collectors "exotes," and makes the point that "The boundaries of individual cultural freedom are always determined by the collective memory of the ways things have been done—of the possibilities for self-expression, the consensual identity bank with slots for various types of rebels as well as conformists" (2008, 352). Vaher says that such collectors "transform their own identity by inhabiting the imaginary utopian territory of 'incredibly strange' sounds and shared cultural history by dropping the needle on forgotten or un-experienced vinyl memories, are just the extreme example of what most of the adventurous record collectors of any kind are doing" (2008, 352).

Kevin M. Moist notes that as in other forms of collecting, such collectors also participate in the expansion of the discourse of specific genres, saying, "[t]he role of collectors has also been significant in relation to specific musical genres, such as blues, by selecting and defining what recordings become established as the essence of the tradition" (Moist 2013, 232). In his book *Retromania: Pop Culture's Addiction to Its Own Past*, Simon Reynolds calls record collectors "renegades against the irreversibility of pop-time's flow, taking a stand against the way styles go out of fashion or run out of steam" (2011, 93). One collector asked about why he collects for this study replied, "I've learned to enjoy more music as I get older and to be more open minded/understanding. Part of the fun of collecting is learning about new things and learning how all things ultimately are tied to each other one way or another" (A. McCubbin, personal communication, January 6, 2015). The social activities of these collectors and the subcultures they participate in creating also assist in the creation of new genres. If, as the Benjamin quote suggests, collecting in general is about negotiating between order and disorder, the discursive practices drive and explain vinyl record collecting's attempt to bring order to the collecting process itself.

As Kenneth Burke observes, however, the language of collecting shapes the range of observations that can be made about these objects. The question of "value," whether use value or exchange value, serves to deflect attention, perhaps, from other issues that might be worth mentioning. Cavell quotes Benjamin's "Theses on the Philosophy of History" as saying, "There is no document of civilization that is not also a document of barbarism" (Cavell 2013, 127). In the work Cavell quotes, Benjamin's discussion of Klee's painting "Angelus Novus" might apply to collectors as well as other perhaps more official historians.

> This is how one pictures the angel of history. His face is turned toward the past. Where we perceive a chain of events, he sees one single catastrophe which keeps piling wreckage and hurls it in front of his feet. The angel would like to stay, awaken the dead, and make whole what has been smashed. But a storm is blowing in from Paradise; it has got caught in his wings with such a violence that the angel can no longer close them. The storm irresistibly propels him into the future to which his back is turned, while the pile of debris before him grows skyward. This storm is what we call progress. (Benjamin 1968, 257)

What sorts of ideological debris get strewn behind in the ways collectors talk about their collections? While Vaher can celebrate exotic record collectors by discussing the two volumes of Vale and Juno's *Incredibly Strange Music*, calling such collectors "active DIY ethnographic surrealists, researchers as much as artists, reinterpreters of collective memory as well as dreamers on their own sonic islands" (2008, 348), he can do so only by dismissing exoticism as a "by-product of Western colonial expansion" (345).

How, then, do record collectors account for the inconvenient questions that might arise out of the collections they amass? What do they talk about, and what do they ignore? How to account for the history of industrial practices and exploitation contained therein? How to account for the thorny question of the way(s) that race and racism contributed to the history of popular music? How to account for the decades of appropriation and the cultural imperialism and hegemony that the twentieth-century music business *also* represents? Tim Gracyk, for example, in his website "Tim's Phonographs and Old Records" discusses in worthy detail all sorts of ephemera of early recorded music, wax cylinders and shellac 78s. Gracyk includes a discussion titled "Music that Americans Loved 100 Years Ago—Tin Pan Alley, Broadway Show Tunes, Ragtime, and Sousa Marches." In a section he calls "Basic Categories," he presents a taxonomy of types of music, in which Gracyk includes parenthetically to ragtime, "(and related 'coon songs')" (Gracyk 2006). Gracyk's discussion of ragtime includes coon songs, as he insists it must, but he is apologetic about even broaching the subject, noting, "'coon songs' have become obscure, virtually no recordings being available today. That is not surprising. The term itself is offensive. Lyrics of songs with stereotypes about African Americans as chicken-stealing, razor-toting 'darkies' are offensive" (2006). He also notes that despite the relationship between coon songs and ragtime, "Today when musicologists speak of ragtime, they usually mean a classic ragtime composition as played by a solo pianist" (Gracyk 2006). While Gracyk's desire to downplay such a distasteful aspect of recorded music history might certainly be understandable given ugly racist associations it exposes, the questions raised by the centrality of coon songs to the early history of recorded music perhaps requires a different approach.

In her study of Africa-American performers of coon songs, Patricia R. Schroeder suggests as much, summing up the calls of other scholars for a more central role for minstrelsy and coon songs in the history of popular music, saying,

> For scholars like these, coon songs thus deserve study not just as a transitional moment between minstrel shows and jazz or blues, but in their own right, as indispensable to our understanding of an American popular music, the larger cultural forces that produce it, and the sophisticated racial commentary of the talented artists who crafted and performed the songs. (Schroeder 2010)

Even when the songs were sung by White minstrels in blackface, however, the foundational role played by the coon song in the history of recorded music cries out for acknowledgment, though. Susan Willis notes that "In any society defined by social inequality, culture is a terrain of struggle" (Willis

1994). Willis says of claims that the struggle to reclaim minstrel culture by African-Americans is not worthwhile,

> the struggle is not simply to reclaim cultural reference points but to work through the complicated relationship between white and black cultures as these have been articulated in mass form and structured by the politics of domination, exploitation, and at times, subversion. In any society defined by social inequality, culture is a terrain of struggle. (1994, 184)

Willis's discussion of efforts to reclaim the cartoon character Mickey Mouse's roots in minstrelsy provides an interesting analog to record collectors of the early era and beyond. She argues that Mickey Mouse's first appearance in the cartoon "Steamboat Willie," in which he was seen "dancing a jig and singing and whistling to 'Turkey in the Straw'" marks him as black and places him in the minstrel tradition, despite the fact that many people today would probably see him as white (Willis 1994, 184). She says of such efforts, "[s]een from this perspective, cultural literacy supplies a tool for unlocking the commodity and revealing the suppressed traces of its counterhistories" (Willis 1994, 185). These suppressed traces and counterhistories assume an ethical imperative in the language of record collectors, shellac or vinyl.

What of the claim, however, that, as Willis states, critiquing the commodities that culture provides is an "exercise in futility" (Willis 1994, 185)? A Marxist, after all, might well say that as commodities, these objects cannot reveal their social meanings, because as commodities they are by definition alienated from the social? David Banash stakes out a place for collectors that puts them at least partially outside of the realm of consumption. Collecting is oriented toward the past and preservation, according to Banash, as opposed to consumption, which is future oriented and destroys as it uses (Banash 2013, 56-58). David Hayes notes the difference in relation specifically to vinyl record collectors in his study, "'Take Those Old Records off the Shelf': Youth and Music Consumption in the Postmodern Age," saying of young people who collect vinyl records because they reject the disposability of current popular music,

> through their retrogressive tastes and practices, these youth effectively disrupt the music industry's efforts to define and regulate their consumer identities, thus restoring a degree of autonomy to an economic relation widely perceived to be over-determined by corporate objectives, youth-oriented marketing campaigns, legal action and other forms of control advocated by the Recording Industry Association of America (RIAA). (2006, 52)

What, however, of the residue of capitalism and ideology? Of the records the collectors he studied collect, Hayes says that they in fact deliberately "ignore these recordings' indelible ties to the music industry's capitalistic framework" (Hayes 2006, 53). Hayes says that these collectors refuse the consumer identities that the music industry thrusts on them, instead "refusing the present and the commonplace in favor of the past and the obscure" (2006, 53). And, however willful the act of opposition might be, whether or not these collectors can completely shake off the dust of industrial capitalism from the objects it provided in their acts of recontextualization is very much an open question.

At the intersection of opposition and obscurity, such collectors create perhaps a counter discourse to the official discourses of the music industry, out of the very objects of industrial capitalism. Michel Foucault has said that "It may be that Marx and Freud cannot satisfy our desire for understanding this enigmatic thing which we call power, which is at once visible and invisible, present and hidden, ubiquitous" (Foucault 1977, 213). Foucault explains,

> There are no relations of power without resistances; the latter are all the more real and effective because they are formed right at the point where relations of power are exercised; resistance to power does not have to come from elsewhere to be real, nor is it inexorably frustrated through being the compatriot of power. It exists all the more by being in the same place as power. (1980)

The languages deployed by record collectors might be said to provide the kind of site of resistance to record industry discourse. One recent example of such resistance by record collectors might be illustrative of how counter discursive practices can work to subvert industrial-capitalist discourse. In response to the question of one of their readers, *Record Collector* includes a story about an "Anglo-Welsh band" called "Mirage" (Shirley 2015). Mirage released only one song during its time as a working band, called "Blind Fury," on a 1980s compilation called *Notepad Productions Volume 1*. Mirage recorded only three songs total before their lack of success ended their career, and only "Blind Fury" saw release at the time. The only reason "Blind Fury" was released at all was that the group paid the record company to press it. Notepad Productions charged each band £200 for fifty copies of the record in order to participate (Shirley 2015).

Remarkably, the release of that record was not the last that Mirage was heard from. In the 1990s, the Notepad compilation came to the attention of Malc Macmillan, who included a "glowing" review of the song in his *NWOBHM Encyclopedia*, published in 2001, a book that has since become a benchmark for fans of the genre. NWOBHM stands for "New Wave of British Heavy Metal," a genre that itself grew out of localized movements in England in the late 1970s and early 1980s, in direct opposition to prevailing fashions and the will of the British music industry at the time. In fact, as a movement and eventually as a genre, the NWOBHM grew out of the punk DIY ethos, and it spawned several independent record labels of its own before being swallowed up by the majors (Rivadavia 2014). Macmillan's reappraisal of the Mirage track has led not just to the increasing value of the Notepad compilation (£30 as of January 2015, according to *Record Collector*'s estimate), but also to bringing the group back to the attention of the music collecting public in general, a fact of which not even the band themselves had been aware until recently (Shirley 2015). In 2015, Mirage is set to release a single based on recordings they made back in the 1980s. The original impulse behind the release of "Blind Fury" may very well have been the exploitation of young hopeful rock musicians, but the song lives on in an entirely different register as an exemplar of a genre that initially grew in opposition to industry sanction. NWOBHM collectors present a sort of bottom up discourse the practices of which, even while they partake of the cast-off products of the music industry, subvert the meanings such products may have originally intended and transform them at least partially into something with an entirely different value. In such oppositional practices, these collectors reflect the influence of punk in the last part of the twentieth century.

Another example of this kind of (re)appropriation by collectors, perhaps on the other end of the spectrum of capitalist exploitation, is *Mojo* magazine's reissue of the year for 2014, *L'Amore* by Lewis (Sullivan 2015). Light in the Attic Records (LITA), an independent label based in Seattle, Washington, are responsible for many of these reissues, including the revival of the career of Sixto Rodgriguez (LITA 2015). LITA characterizes their approach as "commitment to quality, as well as its disdain for convention," and they operate an entire family of independent labels, issuing music on vinyl, CD, and digital download. The Lewis record was recorded and pressed privately back in 1983, but the LP was discovered by collectors from Canada who posted tracks online in 2007 (Sullivan 2015). The original recording and pressing of the Lewis LP was as mysterious as the identity and current whereabouts of the artist who made it. LITA released the record in 2014 to great acclaim, placing the royalties owed to Lewis in escrow in case he was ever found. Matt Sullivan from LITA did some research and discovered that the artist's real name is Randall A. Wulff, and with some trouble Sullivan located him Canada. At their meeting, Sullivan offered Wulff the royalty check for the Lewis record, but Wulff refused the money. Sullivan handed Wulff a CD copy of the record, Wulff handed it back, saying "That was a long time ago" (Sullivan 2015). Wulff leaves Sullivan stupefied by his reaction, wishing him and his label "all the best." Sullivan concludes the story by saying, "we've decided to stop manufacturing his albums and continue putting his royalties aside; it doesn't feel right to profit from his music when he's not involved" (Sullivan 2015). While a record label was involved in the distribution of *L'Amore*, the record's discovery was driven by the passions of collectors and the availability of the Internet as a means of spreading those passions.

Referencing Riesman, Evan Eisenberg makes the significant point that "true" record collectors can be seen as "heroes of consumption" (2005, 15). He calls the heroism of such collectors "closer to the ancient model," because "it is exploit, the dauntless overcoming of obstacles in pursuit of the prize. The prize itself is secondary to the pursuit" (Eisenberg 2005). Paradoxically, the true hero of consumption is also "a rebel against consumption," repudiating acquisition by taking it "to an ascetic extreme" (Eisenberg 2005). Hayes says of his young subjects that they "construct consumer identities that challenge the music industry's attempts to define and regulate behavior" (Hayes 2006, 53). Adorno's point, however, about the reified fetish character of recorded music is perhaps too easily dismissed. As products of industrial capitalism, records do retain something of their commodity character. When Adorno notes that "[m]usic, with all the attributes of the ethereal and sublime which are generously afforded it, serves in America today as an advertisement for commodities which one must acquire in order to hear music," and that the use value of music has been "replaced by pure exchange value," he may not be speaking for the listening public, but he is speaking for the record industry, which has a long history of corruption and exploitation. Will Straw says, "the cultural commodity is fragile because of the discrepancy between the pre-capitalist values it is meant to embody (those of individual freedom or a unified experience) and the standardized, fragmented capitalist forms in which it is produced and received" (1999–2000, 150). And Simon Napier-Bell clarifies in *Ta-Ra-Ra-Boom-De-Ay*, his droll history of the music business, since its beginning in music publishing, "[w]hat pushed the industry forward was not so much the public's love of music, or the musicians' love of playing and writing it, but the publishers' desire to make money from it" (Napier-Bell 2014). Countering the ideological traces built into the products of the music business takes a conscious, perhaps even a heroic, act of will. And, what are we to do with the fact that in the age of digital downloads and streaming media, for many of the intended consumers of the products of the recording industry, the products no longer even bear the marks of Adorno's debased commodities? Those products no longer bear even physical form, being digital "tracks" that can be streamed on Spotify, Pandora, or any one of dozens of other streaming music services or downloaded via torrent for free—can we therefore even discuss them in terms of exchange or use value at all? How can you reify something that takes no physical form?

Is the return to physical media in the face of such technological change the most meaningful possibility of resistance? David Hayes, who it must be remembered published his study of young vinyl collectors before the explosion of streaming music, of the young vinyl collectors of his study, concludes that

> involvement in the culture of vinyl—from hunting down an obscure LP to cleaning it, playing it, listening to it (and negotiating its pops and crackles), archiving it, and, perhaps most importantly, talking about it with other similarly minded individuals—has enabled these young people to operate with a reinvigorated sense of agency in an arena of cultural production and consumption largely overdetermined by corporate interests. (Hayes 2006, 67)

According to Evan Eisenberg, resistance to its nature as an object of industrial capitalism is built into the very grooves of the record—once music was inscribed in shellac or vinyl, he says, "music was now an object that could be owned by the individual and used at his own convenience" (Eisenberg 2005, 24). Of course, such "ownership" is ultimately illusory. According to copyright law, the owner of a sound recording, whether on vinyl or on CD, owns only the object that contains the recording, not the sound recording itself (The Law 2015). Mark Katz says that it's the very portability of the recorded object that contributes to the creation of this illusion. In response to Benjamin's concern for mass-produced music's loss of aura, Katz responds, "[t]rue, mass-reproduced art does lack temporal and physical uniqueness, yet reproductions, no longer bound to the circumstances of their creation, may encourage new experiences and generate new traditions, wherever they happen to be" (2004, 15). Katz cites DJ turntablism, by which DJs manipulate the turntable to create new sounds using the recorded sounds contained in the grooves of the recordings, as just one example of such a new experience (2004, 115). In addition, the RIAA's decades-long battle with both tapers and downloaders might just demonstrate the futility of anyone claiming ownership of such an object so ontologically strange, being both an object and a performance contained by that object.

Returning to Kenneth Burke, it could be said that record collectors create their own terministic screen in opposition to the terministic screen the recording industry attempts to impose upon the consumers of their products. The oppositional discourse of punk to the dominant discourse of the mainstream music industry of the time has certainly inflected this terministic screen. The availability of discourses resistant to reification is perhaps related to what Lukács and later Habermas referred to as "lifeworld," the informal world of day-to-day interactions that resists the rationalization industrial capitalism imposes. Deborah Cook summarizes Habermas's position thus,

> the functionalist rationality of the economic and political subsystems is restricted by its very one-dimensionality. It confronts a "unity of rationality" that lies "beneath the husk" of everyday practice. The univalent rationality of the subsystems conflicts with the multivalent communicative rationality that characterises action in the lifeworld. Although it can undoubtedly be damaged by them, communicative rationality inherently resists the colonising incursions of functionalist systems. (Cook 2005, 56)

While Cook ultimately questions Habermas's "claims about the inherent limits to reification," she concedes at least partially his point about the "multivalent communicative rationality" of the lifeworld. This multivalence perhaps allows for oppositional discursive practices to emerge. Bakhtin refers to this multivalence as "heteroglossia," about which he said, "[l]anguage is not a neutral medium that passes freely and easily into the private property of the speaker's intentions; it is populated—overpopulated—with the intentions of others. Expropriating it, forcing it to submit to one's own intentions and accents, is a difficult and complicated process" (Bakhtin 1981). Like Bakhtin's point about the entry of dialects into literature, the entry of the dialects that are the products of the manufacture and marketing of recorded music into the culture of consumers and listeners both inflects and is inflected by the process, thus making available perhaps other, non-sanctioned meanings to record collectors, opening up the act of mere capitalist consumption to resistance. As Hayes says of his young collectors, "hunting for obscure recordings doubles as an act of resistance against the music industry's maintenance of a constructed popular music narrative" (Hayes 2006, 64). The fact of the matter is, however, that despite their relative obscurity in the present, those records were originally the products of the very industry that is being resisted in the act of hunting them down.

Punk may very well have positioned itself in opposition to mainstream discourse, but despite its emphasis on independence, its oppositional discourse was itself always already embedded in the discourses of industrial capitalism. Robert Garnett says that "[t]he last thing punk, at its best, was interested in was an 'aesthetics of consumption,' and looking back on it from the context of the present, it is surprising to see how Adornoesque it was in its attitude towards the popular culture industry" (Garnett 1999, 21). Surprising because, in the words of Greil Marcus quoted by Garnett, the Frankfurt School critique of the music industry exploded "out of a spot no one in the Frankfurt School, not Adorno, Herbert Marcuse or Walter Benjamin had ever recognised: mass culture's pop-cult heart" (Garnett 1999). Adorno famously said of the products of the culture industry that while in the past they might have served a dual purpose of both profit and "autonomous essence," under modern capitalism, they "are no longer *also* commodities, they are commodities through and through" (Adorno 2002). Punk Rock as a genre sought to reclaim this essence and this autonomy through its oppositional stance and also through its embrace of marginalized subcultures, and it is this oppositional stance that inflects the language of record collectors in 2015.

The use of the term "vinyl" itself to describe records can be seen as the adaptation of industrial language to consumption and collecting. To call them "records" signifies the performance these objects contain, however, "vinyl" denotes the industrial materials of which the object is made. And, while "vinyl" was used as a descriptive term for the objects as far back as the

1950s, it didn't come into wider use until the introduction of the compact disc in the 1980s and '90s as a means of differentiating the two formats. The semiotic slipperiness of "vinyl" (aside from the industrial compound that makes up the object, what, exactly, does it describe?) even leads to confusion and controversy over the proper usage of the term, specifically centering on the plural form of the noun. Younger collectors refer to the objects in their collections as "vinyls," which raises the ire of older collectors and amateur grammarians of all stripes. Interestingly, though, correcting these neophytes does not resolve the issue of the proper plural for the word. Mark Liberman calls the controversy over the use of the plural, "a fascinating case of peeve emergence," by which people voice their annoyance with the way others use language (2012).

Liberman quotes several examples of peevishness over the use of the plural "vinyls" in his blog entry dated June 12, 2012, including this one from a responder to an IP Board forum topic titled "Amputechture vinyls [*sic*]," who says,

> Man, I hate to be the school marm but..."Vinyls" is not a word. The plural of vinyl is "vinyl" like deer is the word for multiple deer. Or you could say records. Not trying to be a jerk, just educating. I've been on some forums where people are tarred & feathered for saying "vinyls." (Liberman 2012)

Liberman makes the point that helpful educators such as this are actually defending a rule that is "doubtful at best." Liberman cites Arnold Zwicky's term "countification," "whereby the plural form of a mass noun can be used to refer to more than one type or instance of the named category of stuff" (Liberman 2012). It is worth looking at the way that Zwicky discusses this distinction, because it illuminates why the way the word "vinyl" is used contributes to making it so slippery. Zwicky says,

> English nouns fall into two classes, C and M, according to their morphological and syntactic properties: C nouns like SHRUB have plural forms, can occur with the article *a(n)*, etc., while M nouns like SHRUBBERY lack plural forms, do not occur with *a(n)*, etc. C nouns typically denote "things" (which can be counted, hence the customary label *count noun*), while M nouns typically denote "stuff" (often an indivisible substance, hence the customary label *mass noun*). (Zwicky 2008)

Through the language of collectors, the "stuff" of industrial production (polyvinyl chloride) comes to designate the "things" that are collected. Liberman concludes, "So 'the plural of *vinyl* is "vinyl"' is an invented 'rule,' more or less the opposite of the general patterns in the language" (Liberman 2012). This raises the question of why the investment in correcting collectors who perhaps naively include the plural "s" at the end of the word. Perhaps Liberman's tongue-in-cheek use of the term "millennial hipsters" to describe young vinyl collectors might provide a clue (2012). The (mis)use of the inflectional ending for the plural is believed to mark the users as new to the experience of collecting, and the inexperience it connotes seems to serve an almost liminal function, placing those who use it thus between the spaces of non- and experienced collectors.

The non-sanctioned counter-discursive practices of outlaw collectors preceded punk rock, and influenced by punk, these discourses largely inflect the practices of both record collectors in the age of social media and the embattled record industry itself. Roy Shuker notes that the industry acknowledged and to some degree catered to the practices of collectors from very early on in its history. Shuker says,

> a range of marketing practices emerged amongst the early recording companies. While these were oriented towards the general market, most also targeted the music collector, several almost exclusively so. The most obvious marketing strategy was the production of catalogues and other advertising material. (Shuker 2010, 22)

The industry did not mind marketing to collectors, as long as they could maintain tight economic and artistic control of the objects they collected. In rock era, however, the issuing of non-sanctioned "bootleg" recordings of popular artists presented a serious challenge to that control. It should be stated that the recording artists themselves also opposed the issue of such bootleg recordings, but, while money was a factor, they did so more because of the challenge these recordings presented to the artists' creative autonomy and to their popular images. While there was a market for bootleg recordings prior to the rock-and-roll era, it didn't really explode until the end of the 1960s, with the release of the first rock bootleg, *The Great White Wonder* by Bob Dylan. The release of this record and of Dylan and The Band's 1967 post-motorcycle-accident recordings helped create what *Record Collector News* calls, "an entirely new segment of the music business and record collector world" (Kubernik 2014). *Record Collector* says that the story of *The Basement Tapes* "charts the beginning of the whole bootleg era" (Humphries 2015, 73). And, while the recording industry fought the release of these illicit recordings through all the legal means at their disposal, they also responded by officially releasing their own version in 1975, called *The Basement Tapes* to underline their raw unfinished nature.

As in the issue of piracy, however, the question of record industry control in the case of contraband recordings turns out to be slippery at best. Also similar to their discourse on piracy, the record labels usually justify their desire for such economic control in terms of "quality" and "artistic integrity," neither claim of which particularly stands up to scrutiny. Clinton Heylin quotes Steven D'Onofrio of the RIAA piracy division as saying, "[t]here are artists out there that are very concerned about bootlegging . . . They hear their voice not as flattering as it is in the studio, being recorded and released permanently, and they don't necessarily like that" (1996, 393). Heylin makes a sort of case study of Dylan, to whom Heylin points as particularly paranoid about bootlegs of his work, noting that it is partly Dylan's own "profligacy" in documenting himself in sound that makes the work of the bootleggers so easy (1996, 394). Heylin numbers Dylan among the artists "who have been the most fanatical in attempting to quash bootlegs," artists "whose judgment [about their recorded legacy] has been called into question the most" (1996, 394). And while *The Basement Tapes* may have been a belated response to the non-sanctioned release of *The Great White Wonder* and other Dylan bootlegs, the final product perhaps points to a flaw in such attempts by both the artists and the labels, a certain lack of objectivity compounded by power-mad shortsightedness. Heylin joins many others in calling the results "appalling," and says it is "perhaps the best argument that archival releases should be left to bootleggers" (1996, 395). Heylin continues, "[t]he Basement Tapes represents the most extreme travesty, mixing stereo tapes to mono, including eight superfluous Band cuts recorded elsewhere and later, and leaving at least as many classic Dylan originals off the double set as were included (1996, 395). This "travesty" was not corrected until Dylan and his label sanctioned the release of some of his archive recordings as *The Bootleg Series* starting in 1991. Release in 2014, the eleventh volume of the series is a six disc set called *The Complete Basement Tapes*.

In his review of *Volume 11*, Harvey Kubernik talks about the disappointment of fans, including "part-time Dylanologist and full time admirer" Gary Pig Gold, with the inauthenticity of the 1975 release (Kubernik 2014). Kubernik quotes Gold as saying,

> It wasn't until the dreaded mid-Eighties, thanks to a then flourishing network of, um, tape traders who were stepping up to do what the music industry couldn't be bothered to, that I finally began to find, hear, collect and duly treasure THE REAL THING. As in those majestically raw-boned, un-tinkered-with, expertly Garth Hudson-engineered, subterranean West Saugerties jewels. Dozens upon dozens of them, in fact! Some funny, some smart; some long, others quick, all basically indescribable… especially when placed within their proper historical—as in Sgt. Pepper's Lonely Hearts Club Band perspective. (Kubernik 2014)

That the tapes were released at all taking an "official bootleg" form speaks much about the influence of the counter-discursive practices of collectors on the market for sanctioned recordings. In response to the dissatisfaction with the 1975 release, "[t]he decision was made to present The Basement Tapes Com-plete [*sic*] as intact as possible. Also, unlike the official 1975 The

Basement Tapes product, these performances are presented as close as possible to the way they were originally recorded and sounded back in the summer of 1967" (Kubernik 2014). The attempts of both artist and label to maintain control of these recordings and therefore Dylan's legacy were to some degree surrendered to the demands of the collectors, revealing what Kubernik calls "a whole shadowy subterranean alternative career for Dylan," which is precisely the thing that made Dylan and his record label so nervous in the first place.

The Beach Boys join Bob Dylan, Led Zeppelin, the Rolling Stones, and the Beatles as rock-era acts that are frequently bootlegged. The belated release of the long-lost 1967 Beach Boys album *SMiLE* also bespeaks the influence of bootleg collectors. In the years since the original release was scrapped, according to Eric Klinger, "rock nerds the world over became obsessed with what *SMiLE* could have been. Bootleg versions were passed around by fans who cobbled together their own understanding of what the album would have been, for an actual track list had never been released" (Mendelsohn and Klinger 2014). Brian Wilson, the creative genius largely considered responsible for the success of the Beach Boys, not to mention the man responsible for the original recording of *SMiLE* getting scuttled in the first place, responded to the demands of obsessed collectors by releasing a rerecorded version of his work as a solo project in 2004, a move that only led to further expressions of dissatisfaction, again because the artist's attempts to control the shape of the "official" release conflicted with the expectations of collectors who had been producing their own versions in ensuing years. Because it did not do enough to satisfy demand, *Brian Wilson Presents SMiLE* was followed in 2011 by the first official release of the official Beach Boys version, compiled by Capital Records using the tapes from 1966 and 1967 that remained in the vault. One of the compilers of the album, Mark Linett talked to *Billboard* magazine when he was still submerged in crafting the final product, and he speaks as one who is aware that *SMiLE* is "one of the most bootlegged records of all time" (Christman 2011).

In fact, the bootleggers seem to be a presence in many of the decisions the compilers made about the project. Linett discusses an archive project he is working on with the band, adding about *SMiLE*, "[a]nd the other important thing is bootleggers tend to present every single take . . . We are obviously going to use the best versions and there are things that we can do that was just technologically impossible when those bootlegs were made in the 1980's [*sic*]" (Christman 2011). In addition to the official release of the album on vinyl and CD, a multi-disc set of the sessions that bootleggers had been trading and using to compile their own versions of *SMiLE* called *The SMiLE Sessions* was released. Ultimately, though, even these releases did not wrest control of this legendary record from the collectors. Reviewing the release in *All Music*, John Bush says, "having a full *SMiLE* album in mono and a collection of sessions in stereo immediately positions The *SMiLE* Sessions as something less than a true bootleg beater—which will undoubtedly lead fans back to extra-legal means" (Bush 2011). Like Bob Dylan and his record label, the best efforts of Brian Wilson and his label to control the discourse surrounding his career inevitably leads to further counter-discourse from the collectors.

Of course, perhaps ironically, it's entirely possible that Dylan's career would have taken an entirely different turn were it not for the work of an earlier collector, Harry Smith. Smith was a collector of obscure early twentieth-century 78s of American blues and country music that he collected in vinyl LP form in 1952 as the *Anthology of American Folk Music*, a move that Mark Richardson says "kickstarted the national interest in folk that would eventually shape the consciousness of the 1960s" (2014). On the *All Music* website, John Bush says that the recordings presented on the album mostly had not been heard for over twenty years at the time of its release, and "it proved a revelation to a new group of folkies, from Pete Seeger to John Fahey to Bob Dylan" (1997). In his review of the CD reissue in 1997, Robert Christgau says that

> for the many young people whose lives were changed by the Anthology, its eccentric virtuosity and arcane historical content constituted a thrilling and startling revelation. This revelation would fuel the coming "folk revival" from the Kingston Trio to Bob Dylan and directly impact such '60s rockers as Neil Young and Jerry Garcia. It would inspire young explorers to scavenge the South for 78s and locate such living musicians as Anthology mainstays Mississippi John Hurt, Furry Lewis, Dock Boggs, and Bascom Lamar Lunsford. The bluegrass style that Bill Monroe invented in the mid-'30s spread north because Smith planted the seeds. (Christgau 1997)

Bush also makes the point that the original 1952 release on Folkways Records was itself only "quasi-legal" in the first place (1997). In fact, according to the liner notes, the recordings on the *Anthology* were not fully licensed until the 1997 CD reissue of the recordings (Notes 1997). The counter-discursive practices of collectors, in fact emerged almost at the same time as the discourse of the recording industry itself. The record producers published catalogs of their recordings to alert collectors to what was available, but as Shuker says, "consequently, collectors often lobbied companies to release recordings of particular repertoires/performers" (2010, 21). Try to control the discourse of popular music as much as it wants, the nature of the products of that discourse is such that it eludes the grasp of the recording industry, and the counter-discursive practices of record collectors continually reassert themselves.

The material that bootlegs make public could be considered to be parts of the refuse of the industrial process. Buried beneath the sanctioned discourse of "performance," "fidelity," and "artistic integrity" lies the discourse of the industrial production processes that brought these objects into being. It is these discourses of production that form the Dead Sea Scrolls for collectors. For instance, in his discussion of an early collector's guide written by Morton Moses, Roy Shuker points out that collectors such as Moses were already gathering and disseminating this information, noting, "that the would-be collector of rare records must learn to distinguish between five types of Victor labels" (2010, 26). In an industry already grown large and complex, label and font variations helped manufacturers keep track of when and where records were pressed. Collectors made use of this information to help them locate rare and valued pressings and to distinguish early from later pressings of a particular recording. It was up to collectors to gather this information as well, because, "early recording companies were frequently very unsystematic in their operating practices, especially in terms of tracking inventory and cataloguing releases" (Shuker 2010, 27). In fact, according to Shuker this laxity on the part of the early industry lead to the importance of compiling discographies as part of the practice of record collecting.

Some early collectors such as John Hammond and Jerry Wexler went on to found or run record labels themselves (Shuker 2010). Sometimes, as in the case of Harry Smith, a particular collector gathers so much of information on the records he or she collects that the collector becomes more of an expert about the recordings than the record companies or the artists. Perhaps one such example is Bruce Spizer, an acknowledged expert of the recordings of the Beatles, particularly those released on their US labels. Spizer has compiled and written several lavishly illustrated comprehensive guides to the Beatles' US and British recordings, and he participated in the reissue on CD of two volumes of their Capitol Records catalog. Due to differences in publishing rules and industry practices, the Beatles records had originally been issued in different configurations in the U.S. and the U.K. When their recordings were issued on CD in the beginning of 1987, their record company decided to standardize the Beatles' catalog around the world to the U.K. versions (Erlewine 2004). And, the differences between the original UK and US versions went beyond sequencing and track selection. Stephen Erlewine notes that Dave Dexter, the Capitol Records executive responsible for the Beatles' output in the U.S. in the 1960s, changed the sound as well, and that "the original mixes were given ludicrous layers of echo on the stereo versions that changed the feel of the albums" (Erlewine 2004). Despite the reservations of Apple, the Beatles' record label, the demands of American collectors led to the release of the two volumes on CD in replica sleeves based on the original US master tapes in 2004 and 2006. Spizer himself contributed the liner notes to the second volume (Spizer 2006).

It has been noted that "early recording companies were frequently very unsystematic in their operating practices, especially in terms of tracking inventory and cataloguing releases" (Shuker 2010, 27). And, while record keeping practices may have improved with time, the complexity of the endeavor of manufacturing and distributing records was still such that it was extremely treacherous going to try to appraise a particular pressing. Spizer needed several books just to cover all of the variations of the Beatles' original releases in the U.S., two volumes to cover the Capitol label and two further volumes devoted to the labels that licensed their U.K. recordings prior to Capitol's decision to issue them in the U.S. in 1964. Spizer attempts the very difficult feat of cataloging in order the Beatles Capitol pressings on 45 and LP in release order including all variations of labels and sleeves, variations that are themselves a result of the records being pressed and the sleeves printed at different times and in different locations (Spizer 2006, ix–x). These regular industrial practices were compounded by the explosive success of the Beatles in the U.S., meaning that the records were selling so well that they required multiple pressings and that they were sometimes reissued at a later date.

While Capitol Records may have provided catalog numbers to assist collectors in their searches, through close examination of the objects and information exchanged by collectors through sources both sanctioned and unsanctioned, including discographies and fanzines, a collector such as Spizer collates information on the pressing plant that pressed the record, the typesetting variations on the record labels, the types of sleeve design, the places where the sleeves were printed, whether or not the record was a promotional copy distributed to DJs, and even the subcontractors that Capitol used when their capacity to press Beatles records was exceeded by demand (Spizer, Part One 2000, xi). In his extended discussion of Capitol Records' pressing plants, to cite one minor example, Spizer notes that "first pressings of Rubber Soul do not have 'THE BEATLES' on the label. This oversight was corrected on subsequent pressings" (Spizer, Part Two 2000, 236). He further informs Beatles collectors that "[o]ne can determine where the albums were manufactured by examining the label and trail off areas" (Spizer, Part Two 2000). While Spizer's vast knowledge of such Beatles manufacturing ephemera has given him a foot inside of the industry, the Internet has made possible an explosion of similar archeologists, while blogs, torrents, and streaming sites have made possible the proliferation of such Beatles archeologists.

One such example is called *The Beatles Rarity*, a website published by Happy Nat (http://www.thebeatlesrarity.com/). Happy Nat uses his blog posts to talk about Beatles rarities, and he includes a post he calls "Rarity of the Week," in which he streams a particular rarity for his readers to hear. The site also includes a "Collector's Corner" in which Nat shares details of rare pressings that he owns, including several original copies of the notorious "butcher cover" of *Yesterday . . . and Today* (Nat 2013). Happy Nat also takes reader questions, not only on his website, but also on Facebook and

Twitter, and he answers them in a blog post he calls "#asknat," and he includes a searchable database of his collection. Such independent curation flies in the face of the consumerist discursive practices of the recording industry, which, even while it sometimes directly appeals to the collectors, mainly values the immediate and the disposable over the "obscure" and the "authentic." David Hayes says of such practices that they

> disrupt the music industry's efforts to define and regulate their consumer identities, thus restoring a degree of autonomy to an economic relation widely perceived to be over-determined by corporate objectives, youth-oriented marketing campaigns, legal action and other forms of control advocated by the Recording Industry Association of America (RIAA). (2006, 3)

Will Straw defines this act of resistance specifically in relation to time when he says, "If cultural-historical time is experienced as the constant eclipsing of texts and meanings, this may have as much to do with the effect of consumption in diminishing a cultural commodity's exchange value as with the fragility or hazy nature of cultural use value" (1999–2000, 152). Straw calls the collecting of such knowledge by collectors an "ongoing enterprise of vernacular scholarship," and he notes that the impulse behind such scholarship is "partly about exhaustive documentation and the pacification of historical chaos" (1999–2000, 168). The development of the Internet has seen a correlative proliferation of such vernacular scholarship, along with many thousands of artist discographies and official, semi-official, and unofficial vinyl collectors' guides. The sites *All Music* and *Discogs* attempt to gather such information in a more comprehensive manner, but the nature of music collecting and curating renders such a task Herculean at best.

In fact, it could be argued that through a sort of bricolage, record collectors create something more akin to a pre-industrial, mercantile economy using the detritus of industrial capitalism. In his work "Bricoleur and Bricolage: From Metaphor to Universal Concept," Christopher Johnson says of the "tools and materials" of the bricoleur,

> These elements are, so to speak, multivalent, that is, they retain a certain determinate use value, but because of their abstraction from their original functional context there is a degree of manoeuvre, or play, in their redeployment: they are overdetermined in their history but underdetermined as to their potential use. (2012, 362)

While the records themselves may quite literally bear the marks of industrial capitalism, the bricoleur alters and gives them new meanings through acts of recontextualization and the alternative histories compiled in his or her archeology and curation of popular culture. New genres overturn old ones, requiring new systems of value both use and exchange. The arcane and the obscure, previously unvalued or undervalued genres of music, such as "psych," "sunshine pop," "Canterbury scene" and even "Dutch prog" (Heatley 2014) are rediscovered in the dusty corners of the attic of popular culture, the fact that they fell between the cracks on their initial release contributing to their scarcity, and therefore their value.

As stated before, the marks of industrial capitalism make up the data gathered magpie-like by the bricoleurs of vinyl record collecting. While Spizer may be considered a premier American expert of Beatles' collectables, the UK-based publication *Record Collector* started out as a Beatles fanzine. Shuker quotes Johnny Dean, the original publisher as saying of the reason for the shift, "that many of you [the readers] have a deep interest in collecting the records and memorabilia of other important artists and groups as well as the Beatles" (2010, 153). The magazine has not entirely lost its Beatles bias, however, even as it has expanded its mission to cover a wide array of collectable music, though mainly in vinyl format. In 2012, they commissioned Nick Farmer to gather together the most comprehensive guide yet published to the highly collectable first British issue of the album *Please Please Me*, previously discussed in chapter 2. *Record Collector* publishes the *Rare Record Price Guide* on an annual basis, and the reason for this commission was "keen collector" Farmer's desire to correct errors that he had discovered in the guide (2012, 74). Of his task, Farmer says,

> I have been there and back on this project, getting a little bogged down in mother and stamper codes, but subsequently settling on a more straightforward approach to the listing, which is based around label changes . . . However, the stamper codes analysis has proved useful as an accurate measure of relative scarcity for each pressing. (2012, 74)

For his study, Farmer sampled "around 1500 recorded auctions/sales over the last 10 years" (2012, 74). He divides his study into variations in the disc itself, the label, and the sleeve. Farmer goes into detail about the process of manufacturing records as a means of identifying what constitutes a first pressing. The information he provides in this discussion gives a newcomer to vinyl collecting a relatively comprehensive primer into both the language and the processes of the industrial production of vinyl records.

Farmer begins defining the terms of the discussion with the master tape, then goes on to define the lacquer, the matrix, the mother disc, and the stampers. The master tape contains the final mix and edit of the record, and from the master tape a disc called a lacquer is cut using a cutting lathe. The matrix is a "male" version of the lacquer made through the process of electroplating, which in turn is electroplated to make the female "mother," which is used to create the stampers, two for each side of the record (Farmer 2012, 76). Each part of this process is coded by the manufacturer, and these codes are stamped or etched into the "deadwax" on the runout groove of the record, visible if it is held up to the light. Farmer devotes a significant portion of his study to decoding all of that deadwax information, while noting that such information existed in the first place in order to enable "EMI to trace any manufacturing flaws to a particular process or tool used" (2012). For instance, Farmer says of the matrix code that it "shows the master tape recordings used to cut the lacquer disc in the studio, usually three letters followed by a two- or three-digit number" and he identifies the code for the first mono master of *Please Please Me* as "XEX 421 for Side 1 and XEX 422 for Side 2" (Farmer 2012, 76). While Farmer's information can claim to be relatively comprehensive, since multiple stampers were being used at the same time, the chances of identifying a true first pressing are difficult at best, that such a pressing can be identified at all seems dependent upon the detective work of bricoleurs such as Farmer.

The *Rare Record Price Guide* is a respected authority on "collectable U.K. releases," providing buyers and sellers with the information they need to value their records (RRPG 2010–2015). A perhaps less official source would be something such as Anorak's Corner, a Japanese site that describes itself as "200+ pages devoted to Rare Soul Music" (U.S. Pressing Plants n.d.). This site is run by a man named David in Tokyo, Japan, and while he sells records on the site, it is mostly devoted to displaying his passionate knowledge of rare soul records from the 1960s. To help readers identify those records, David helpfully includes a relatively comprehensive guide to US pressing plants and label printing companies, as he says to "unravel some of the markings that appear within the 'deadwax' or 'run-out groove' sections on US 60's discs" (US Pressing Plants n.d.). Counter discourse uncovers a further level of counter-discursive practice in Anorak's discussion of a type of re-appropriation that originated with Jamaican sound system DJs and continued into the Northern Soul period in England. Intense competition among sound systems in Jamaica in the 1960s meant that DJs had to be circumspect about the records they were playing, originally blues and soul 78s purchased in the States. They called the records "secret sounds," and, in order to keep the identities of these records secret, DJs scratched the original information off the labels and/or covered it up with false information. In his history of the sound systems, John Constantinides says, "[c]ertain strategies were employed to ensure a record's exclusivity, such as renaming the songs or scratching off the record label" (2002). Anorak provides a list of over 650 such records, including both the false identification and the original covered-up label information. He states that in some cases it takes years to identify the actual performers and labels for these records (Classic Cover Ups 2013). Commercial considerations aside, in the case of such practices use value certainly seems to eclipse exchange value almost completely, which paradoxically contributes to the later exchange value of the records on the secondary market.

In cases such as identifying the cover-ups, the detective work involved in uncovering the true source of the record becomes almost as much the prize for the collector as the object collected itself. For most collectors, however, knowing what one has is only part of what makes up the prized item. The condition of the record also plays a role in its value. No matter how complicated the task of identifying it, having a rare first pressing of *Please Please Me* loses its luster to some degree if it is in less than "mint" condition. Of condition as a determinant of value, *Goldmine* magazine says, "their relative rarity and demand is important, but a collector or dealer will pay much more for a record in Near Mint condition than one in Very Good Minus condition" (Record Grading 2010). *Goldmine* provides the standards for grading the condition of records that are adopted by most collecting buyers and sellers in the USA, offering grades from "Mint" down to "Poor" along with descriptions to match for both the media and the sleeves. For example, "Near Mint" is listed as,

> *NEAR MINT (NM OR M-)* A good description of a NM record is "it looks like it just came from a retail store and it was opened for the first time." In other words, it's nearly perfect. Many dealers won't use a grade higher than this, implying (perhaps correctly) that no record or sleeve is ever truly perfect.
>
> NM records are shiny, with no visible defects. Writing, stickers or other markings cannot appear on the label, nor can any "spindle marks" from someone trying to blindly put the record on the turntable. Major factory defects also must be absent; a record and label obviously pressed off center is not Near Mint. If played, it will do so with no surface noise. (NM records don't have to be "never played"; a record used on an excellent turntable can remain NM after many plays if the disc is properly cared for.)
>
> NM covers are free of creases, ring wear and seam splits of any kind (Record Grading 2010)

Goldmine notes that they provide high standards and that all records must meet these standards to qualify for the grade, no matter how old they might be. They make the perhaps not surprising claim that "no more than 2 to 4 percent of all records remaining from the 1950s and 1960s are truly Near Mint," which is why prices are so high for such records (Record Grading 2010). Thus the magazine provides a relatively solid set of standards that are considered effective and trustworthy enough that their use among collectors goes mostly without remark.

Grading records, however, is no exact science, and even a source as trustworthy as *Goldmine* does not escape without scrutiny and concomitant controversy. Immediately following their post on grading records on their website, the "comments" section calls their standards into question. Under the heading "Very Good Plus (VG+)," *Goldmine* lists "VG++" and "E+" as possible alternatives. The first reader response to the guide comes from "John," who says, "I'm sorry, but really to keep the grading of Lp's [sic] less of a mystery to the average (and not so average) Joe, a VG+ record should be just that. It shouldn't be also named 'Excellent,' or allowed to be denoted with a +, or ++" (Record Grading 2010). John is more generous than the next commenter, called "You've Got To Be Kidding Me," who responds, "Fire this guy, Goldmine. Don't use VG++ Stop already" (Record Grading 2010). "Phuc Yew" would rather use "VG++" than "M-," and "Kyle" questions the fact that "Good" does not meet his own personal standard of "good," and asks why there are not more grades between the grades *Goldmine* provides (Record Grading 2010). Almost any time in any collector's forum the question of the *Goldmine* standards comes up, and it is not infrequently, some or all of these same issues are raised. For instance, in a 2013 thread on the Steve Hoffman website titled "*Record Collector* U.K./*Goldmine* U.S.—Vinyl Grading Systems," user "stenway" responds, "EX and EX+ Dont Exist! stop use it, also to stop use VG++ (double plus dont exist) or Strongest VG+ or things like that [sic]" (Chris R 2003). The desire for certainty, for there to be some intrinsic meaning to the various grades assigned by these semi-official sources is clearly a strong one, and even bringing it up causes comment for ontological assurance in establishing the grades.

Identifying and valuing one's prizes is cause for both effort and anxiety on the part of the collector. The reward, however, is considered to be worth it. Like one collector interviewed for this study who says, "When I returned to vinyl and started collecting again, my huge love for music returned. There is something about the imperfections and the ability to see and touch and not just hear that seems to make a connection with me," the ability to see and touch the music they collect can be a very significant aspect of the collecting impulse (J. Henschel, personal communication, January 26, 2015). Of collecting in general, Baudrillard says that it is "first and foremost a discourse directed at oneself" (1994, 22). He hastens to add however, "there comes a point when the self-absorption of the system is interrupted and the collection is enrolled within some external project or exigency," and he accounts for both the impulse to collect and the anxiety when he says,

> the collector is driven to construct an alternative discourse that is for him entirely amenable, in so far as he is the one who dictates its signifiers—the ultimate signified being, in the final analysis, none other than himself. Yet in this endeavor he is condemned to failure: in imagining he can do without the

social discourse, he fails to appreciate the simple fact that he is transposing its open, objective discontinuity into a closed, subjective discontinuity, such that the idiom he invents forfeits all value for others. (Baudrillard 1994, 23)

Baudrillard asks the question that collectors are "bound to ask," "can objects ever institute themselves as a viable language" (1994)? Can the object so subjectively valued speak its value to others? That it ultimately cannot drives the desire of the collector for some objective form of validation, and sources such as *Record Collector* and *Goldmine* step in to fill this void and alleviate the anxiety it creates. *Goldmine* highlights both the imperative of the effort and its inevitable failure in its "10 Commandments of Record Collecting," deemed necessary because "When it comes to buying and selling records, there is a vast difference between the 'value' a proud owner perceives the record to have and the price it actually will bring at the time of sale" (Sliwicki 2013). One of the "commandments" presented is "Thou shalt not mistake a guide with a holy writ," implying that a holy writ is what is being sought when such a guide is consulted (Sliwicki 2013). Such a thing is not possible, of course, because of the very nature of collecting and of valuation; as *Goldmine* advises, "a price guide listing, an expert's appraisal or one isolated auction result is not to be mistaken as a guarantee of a return on investment for a seller parting with a particular record. The real-world sale price can vary widely from the so-called value market value" (Sliwicki 2013). Even the "10 Commandments" themselves cannot be considered to be "etched on stone tablets," themselves becoming yet another guide to interpreting the *Guide*. Such is the nature of counter-discursive practice that constructs its own system of value in opposition to the record industry discourse that it re-appropriates and to which it sets itself in opposition.

Counter discourse is always subject to cooptation by the machinery of industrial capitalism, one reason, perhaps, for industrial capitalism's continued vitality. No matter how hard the collector tries, the dust of commodity fetishism cannot be entirely shaken off discursive practices that at least partially construct themselves from that very dust. The next chapter will examine the ways that commodity fetishism reasserts itself into the discursive practices of vinyl record collectors, with varying degrees of success resulting from the continuing tension between discourse and counter-discourse. Re-appropriation of and resistance to the dominant discourse among record collectors persists, even as the recording industry attempts to reassert itself at the sites of their collecting practice. It is probably worth noting that the example of counter-discursive practice that this chapter began with, the film *WALL-E*—itself a product of the Disney Corporation—is available on DVD, Blu-ray disc, and download, if not from Buy 'n' Large, then from Wal-Mart, Target, Best Buy, or a megastore near you. Of course, it can be argued that Disney, by its deft management of the properties within its vaults, for decades provided a model to other content providers for how the demands and the expectations of collectors can be manipulated for fun and profit.

Chapter Five

"Cabinets of Wonder" or "Coffins of Disuse?"

Reissues, Box Sets, and Commodity Fetishism

We begin with a tale of two reissues. A March 2015 review of the vinyl reissue of the seven original studio albums by Creedence Clearwater Revival in *Classic Rock* magazine expresses displeasure at a botched job by Universal Music. After listing leader John Fogerty's gifts as "a voice as capable of great savagery as it was of heartbreaking tenderness, an incredible guitarist, and a writer of brilliantly catchy, era-defining songs," the reviewer notes that the band's catalog "deserves a little more love than it has received here" (Lewry 2015). Given the considerable gifts of Fogerty and the band, *Classic Rock* does not leave the reader to wonder how far a "deluxe" reissue package would need to go in order to provide a representation adequate to the task. The reviewer states,

> In the age of the deluxe, forensically curated compendium, The Complete Studio Albums is another disappointment. The original Fantasy labels are in place, but there's no sleevenotes, no booklet, no set of limited-edition art prints, no phial of authentic bayou water. Worse, Pendulum isn't the gatefold version adorned by Baron Wolman's spectacular image of a triumphant John Fogerty at the Oakland Coliseum. (Lewry 2015)

Proclaimed the "Reissue of the Month" by *Record Collector* magazine, a review of Third Man Records' reissue of a second volume of 1928–1932 sides from Paramount Records gives the reader a better indication of how better to love such artifacts of recording history. The review describes the package as an art-deco, blue-velvet-upholstered aluminum phonograph-style box [that] houses six white vinyl albums featuring 87 tracks (with holographic images instead of labels); a 256-page hardcover book chronicling the label's story with art plates; a 400 page *Field Manual* with 175 artist biographies by set co-producer Alex van der Turk; the full run of 90-plus Chicago Defender ads, which inspired R. Crumb; plus mail-order catalogs and memorabilia.

It should also be noted that for the $400.00 list price, the purchase of this package includes a further eight hundred bonus tracks available for digital download. Grayson Currin's review of the first volume of this set in *Pitchfork* compares the packaging to "a family of nested matryoshka dolls" (Currin 2013). The first volume came in a "hinge-and-clasp oak Cabinet," short for "Cabinet of Wonder," which is how Third Man Records describes the object.

Third Man Records is an independent label founded by the artist Jack White originally to issue his records with The White Stripes on vinyl, but the label expanded to include a record store in Nashville and an analog recording studio. Third Man also began issuing and reissuing recordings of other artists on vinyl, and the Paramount sets are part of a joint project with Revenant Records. Among his reasons for reissuing these recordings in such an elaborate vinyl package, White lists, "I wanted to make it as appealing as possible to somebody in a physical, tactile way, all of that, the smell, everything, so as to lead you into the incredible stories that are contained in the music" (Rohter 2013). The largely positive *Pitchfork* review notes of the presentation, "Through scrupulous research, audacious design, and ostentatious packaging, this two-volume collection's first installment does precisely what the best box sets intend to do—add proper deference and context to music that remains vital and significant" (Currin 2013). Of course, as they must, both the *New York Times* article and the *Pitchfork* review single out the $400.00 price tag that *Volume One* shares with *Volume Two* of the Paramount package, *Pitchfork* making specific reference to the "rich old men" who will be able to afford it, and worrying that its release only delays the slide into oblivion of these recordings—Currin worries that they will become "Coffins of Disuse"—and leaves us with the question, "Why save something only to let it sink into sure atrophy?" (Currin 2013). Perhaps the answer to Currin's question resides in the nature of the commodity his review is describing. What sorts of processes contribute to creating the kinds of cultural heritage that Third Man both celebrates and commodifies in this package? What is responsible for the elision of the cultural into the commodity? Indeed, can the "Cabinet of Wonder" and the "Coffin of Disuse" be one and the same?

It should be noted at this point that even without all of the deluxe packaging, in all of its complexities of production and distribution a recording is already comparable to the aforementioned "family of nested matryoshka dolls." The last chapter discusses vinyl record collecting as a form of counter discourse to the discourse of commodity fetishism, a counter discourse that constitutes itself from the very materials and language of the record industry it opposes. The record industry is not without its own tools, however, either for fighting or for co-opting this counter discourse. Louise J. Kaplan describes the forces of industrial capital as "devious," because, "disguising themselves in forms that constantly [shift], [they present] themselves as more powerful than the human rationalities and passions that might be enlisted to oppose them" (Kaplan 2006, 132). In his study *Identifying Consumption*, Robert G. Dunn says, "the complex nature and function of commodities themselves as objects having exchange, use, and other values that give form and meaning to our individual and collective lives as consumers" (2008, 22). As both a *cultural* product and a *product* of culture, the recording is perhaps uniquely positioned to demonstrate the complexities of commodity fetishism, because, as written records of human performances, reification is inscribed into its very grooves. As objects of consumption that contain musical performances, records can perhaps be said to provide the music by which Karl Marx's table can do its dance.[1]

According to Marx, the fetishization of the commodity as a process by which human beings project their own alienation onto the goods they purchase, and he defines the "commodity-form" as, "the definite social relation between men themselves which assumes here, for them, the fantastic form of a relation between things" (Marx 1990, 165). Again, according to Dunn, commodity fetishism "points to a mode of representation in which the circulation of commodities constitutes a set of appearances that distort or falsify the 'real' social relations of production and the 'true' character of labor as the source of all value and as a vehicle of self-activity and expression" (2008, 28). Commodities offer consumers access to a world of social relations from which they are alienated by and through their alienation from the products of their labor. Dunn summarizes the Frankfurt School position on commodity fetishism of the arts saying,

> With the extension of commodity relations to both "high" and "low" realms of culture, individuals are turned into passive consumers of predetermined and packaged experiences. As a consequence, art loses it capacity for transcendence, insight, and redemption, and popular entertainment reproduces the experience of factory labor, providing easily digestible but ultimately regimented, superficial, and boring amusements. (2008, 35)

Worse, this fetishism imprints itself upon the very consciousness of the individual consumer, as industrial capitalism offers a cure to its own ailment, since "[a]s a way of life, capitalism generates feelings of insecurity, anxiety, and dependency, causing individuals to seek relief and compensation in the sphere of consumption through its 'substitute gratifications'" (Dunn 2008,

35). In the process of becoming a commodity, a manufactured object loses the particularity of its mode of production as it becomes a screen onto which consumers project their needs, dreams, wishes, and desires. Dunn quotes Marx as saying, "the product steps outside this social movement (of production, distribution, and exchange) and becomes a direct object and servant of individual need." (2008, 25). It goes without saying that the object consumed also loses its historical particularity, the very object of the archeological excavations of collectors discussed in the last chapter.

Evan Eisenberg calls the record collector "a child of the supermarket," noting, "[h]is self-expression is almost entirely a matter of selecting among packages that someone else designs" (Eisenberg 2005, 166). Bartmanski and Woodward say that the expensive reissue phenomenon described here is "driven by the forces of heritagization within music industries" (2015, 102). It might be worth asking at this point of what the industrial history contained by the Third Man Paramount reissues might consist. Aside from the question of the of labor practices in the manufacturing process, the story of Paramount Records is one of enormous determination and artistic achievement just as much as it is a story of rank opportunism and exploitation. Paramount Records began in 1917 as an offshoot of the Wisconsin Chair Company, manufacturing cheap and lower-quality popular records. It wasn't until 1924 and the purchase of Black Swan Records that Paramount came into contact with J. Mayo Williams that Paramount seriously entered the "race records" business and achieved its greatest success.

Until the 1920s, the Africa-American audience for recorded music was treated as an afterthought, if it was treated at all. While the Black-owned Black Swan Records demonstrated that there was a substantial and profitable audience for records by African-American performers, increased competition from the previously indifferent major labels helped to drive the label into bankruptcy, propitiating its eventual sale to Paramount Records. Mayo Williams, the first African-American record executive to work for a White-owned label, saw opportunity in this setback, because while White-owned labels desired to profit by selling African-American records to an African-American market, according to William Howland Kenney, "[n]one of the record companies wanted to associate openly with either the African Americans who bought or those who made race records" (1999, 126). Williams offered his services as an intermediary, and he was given the task of running Paramount's Chicago recording program, albeit without salary or official title. Such was Williams's position with the label that he was largely kept in the dark about the fact that race records formed the greatest part of Paramount Records' business, and that even though he was not compensated directly by the label, the efforts of his artists were largely keeping the label afloat. Norman Kelley says, "[s]o complete was Williams' isolation from his employers that he did not know that Paramount was primarily a 'race' label, in which he was a principle player" (2005, 95). In fact, Williams himself had antipathy for even categorizing African-American records as "race" records, because he felt that the category was demeaning (Kelley 2005, 97). Williams was largely responsible for scouting and recording talent, but on the corporate level decisions were made on an economic rather than an artistic basis, and the artists Williams found would only be allowed a second recording session if sales on the first "reached ten thousand copies" (Kelley 2005, 97).

While Williams's accomplishments in helping to establish Blues as a legitimate musical form are without question, and while his pioneering attainments as an African-American recording executive deserve recognition, Williams himself understood completely the business in which he operated, the ways in which talent and labor were exploited in the name of profits, nor did he exempt himself from such accusations, even when they were leveled at him. Of his dealings with the artists he recorded, Williams said, "I was better than fifty-percent honest, and in this business that's pretty good" (Kelley 2005, 103). One of the ways in which Williams took advantage of the artists he recorded was through gaining control of their publishing rights. According to William Kenney,

> Paramount had created Chicago Music as a satellite music publishing company that could buy, own, sell, and issue licenses for mechanical rights to the musical selections that Paramount recorded. Paramount made Williams manager of Chicago Music, in which capacity he used various stratagems to wrest copyright or mechanical rights from the performer/composers, and arranged to have songs scored for publication and lead sheets registered for copyright with the Library of Congress. He earned one half of the 2¢ Royalty that went from the record company to whoever owned copyright on the material recorded. (1999, 127)

Since he drew no salary from the record company, such methods of cheating artists were Williams's chief means of making a living. One of the ways in which he succeeded as an African-American in both a business and a culture hostile to non-white males was by keeping a low profile and playing his cards close to his vest as far as his business practices were concerned. Williams held the record executives for whom he hustled, the blues singers who recorded for him, and even his closest employees at arm's length as a matter of course. One of those employees, a talented assistant named Aletha Dickerson, played perhaps part of the role in his own success that he himself had played in Paramount Records' success. According to Norman Kelley, Dickerson "performed the actual clerical chores Williams had been hired to do" (2005, 98). Williams considered Dickerson to be a talented songwriter, but he refused to record her songs, because he wanted to discourage her "from overestimating her own importance" (Kelley 2005). Furthermore, in order to frustrate any ideas she might have about replacing him, he never discussed business with her, and he "took pains to ensure that she would learn nothing about Paramount operations beyond what she could infer from his laconic correspondence" with his superiors at the label (Kelley 2005, 98). The saga of Aletha Dickerson, J. Mayo Williams, and Paramount Records at least partially represents the history of twentieth-century American music, a history of monstrous racial exploitation, at least as much as it is the history of the creation of an influential and lasting art form.

Can such a history be contained in such gorgeous packaging as the Third Man Records' *Cabinet of Wonder* provides? The influence of Jazz and Blues music on the culture of the twentieth century and beyond is incalculable—Kelley makes the not insignificant point that "[w]hen people worldwide say *American* music, they often mean, by way of shorthand, *black* music"—but it is at the same time a product of oppression and exploitation, both of which inscribe themselves into the very grooves of the records we collect. Kelley places the history of Black music within a larger context of social and cultural oppression, saying,

> [t]hrough various modes of production and avenues of exchange, the relationship between the two races has always rested on whites' ability to exploit and dominate blacks' bodies, images, and cultures. In the case of music, blacks have rarely received the just benefits of their work, especially in comparison to their white counterparts and those who control the music industry. (2005, 6–7)

Does the release of a deluxe package such as *The Rise and Fall of Paramount Records, Volumes 1 and 2* help to recover that history, or do such luxury items through their reification of the beautiful object merely seek to reinscribe commodity fetishism in another register? What to make of a cultural object such as the remastered vinyl edition of Tupac Shakur's album *All Eyez on Me*, distributed by the Death Row Koch label and available from Amazon.com for $38.88 (All Eyez 2001)? Norman Kelley uses the continued exploitation of Shakur and his music as a "primary example" of the colonization of black recording artists by major record labels, going as far as citing a *New Yorker* article by Connie Bruck that implicates Shakur's record label in his murder (2005, 8). Entertainment One purchased the assets of Death Row Records in 2006 during bankruptcy proceedings, along with Shakur's master recordings. Shakur's mother is currently in the process of suing the company to obtain the rights to her son's master recordings over unpaid royalties dating back to 1997 (Sterling 2013). Afeni Shakur is quoted as saying, "I believe it is our responsibility to make sure that Tupac's entire body of work is made available for his fans" (Thomas 2013). The question is whether the release of another deluxe package of an album that fans have already paid for, perhaps more than once, is consistent with that responsibility, or whether it is merely another way of exploiting recordings that owe their own existence to the "colonization" of an African-American artist.

As for labor practices in the factories operated by the record companies, in a 1937 issue of *Downbeat* magazine, John Hammond wrote about their treatment of their plant workers. Despite the fact that he worked for Columbia Records at the time, Hammond exposed "deteriorating working conditions" in the plants that Hammond felt had led to "a dramatic decline in the physical quality of records being sold to the public" (Prial 2007). Dunstan Prial says about Hammond's crusade,

> Showing no reservations toward biting the hand that fed him, he targeted a Columbia manufacturing plant in Bridgeport, Connecticut for particular scrutiny. With a keen eye for detail, he described the intense heat and foul smells generated by mixing the melted shellac used at the time to make records. Over time, these oppressive conditions wore down the factory workers, affecting their concentration, he argued. (Prial 2007)

Hammond also repeated stories told to him by former employees of the Columbia plant being "cited for violations of Connecticut labor laws." Furthermore, Hammond noted that RCA Victor was "making the best-sounding records in the industry," at least partially because "its factories had recently been organized by the United Radio and Electrical Workers. Prial says that this article got Hammond his first full-time job in the record industry, when he "hit it off" with Dick Altschuler, Columbia's president at the time, after Altschuler angrily called him in for a meeting regarding his article (2007). Nor did his new insider status last more than one month, reportedly because Hammond quit in sympathy with a secretary who was fired for organizing other secretaries (Prial 2007). The deluxe, 180 gram vinyl reissue package can tell its user something about the history of the music it contains within its grooves, but it must remain silent about the history of the labor that went into its making.

To the degree they can, deluxe reissues participate in the recovery of musical history, but they offer only a partial view of the history they recover. Including facsimile tour programs, concert tickets, and other trinkets serves to obscure that purpose as much as it reveals it, offering mere simulacra in place of honest archeological assessment, to the degree that such a thing is possible. Is such repackaging of music an attempt on the part of record companies to pay homage to our collective musical heritage, or is it merely a cynical attempt to extract yet more money from the artists who have been considered to be steady sellers all along? Two related phenomena point specifically to the latter impulse. First, the record labels release deluxe editions of albums that have barely seen their original release, sometimes as few as six months after. Second, record labels make deals with specific outlets to offer different versions of the same album even in its initial release containing store-specific "bonus tracks," ensuring that the biggest fans of the artist will purchase more than one version of the same record. Of these two phenomena, Andrew Martin says, "it seems like more and more artists (and their labels) are looking to cash in on the idea of an 'exclusive' release or something that's essentially the same album with a new coat of paint" (Martin 2012). In this case, exploitation by the record labels cuts both ways, as the record labels take advantage of fans as well as musicians.

While Martin makes the point that he is not talking about vinyl reissues of heritage recordings, the same cynical impulses can and do govern these releases, and, in what may be seen as an even worse manipulation, they often present themselves as high-quality, audiophile recordings when they are nothing of the sort. In the case of these vinyl reissues, however, the question is not so much one of the extras or bonuses offered—although heavy-grade virgin vinyl might be considered to be one of those extras—so much as it is one of the respect the reissue pays to the "aura" of the original recording. Writing for *Goldmine*, Dave Thompson makes this point, saying, "Asked whether a new vinyl reissue should be pressed from its original analogue masters or if the manufacturer should instead employ the best-quality-possible digital remaster, customers were adamant that only the original—or as close as possible—analogue tapes should be employed" (Thompson 2013). This desire puts collectors at odds with the industry and retailers, who believe that the bonus tracks and digital sound that some vinyl reissues feature is their main selling point, or, as Thompson says, "the more modern bells and whistles attached to a disc, the better its chance of selling to the general public, as opposed to committed vinyl fans" (Thompson 2013). The mercenary motive behind the deluxe reissue package is made plain by Andrew Martin, who says, "Because the sales of re-releases count towards the sales of the original project, labels and artists certainly seem to think so. This is a time in the record industry where every bit of profit has to be squeezed out of a music release" (2012). Nor does the profit motive affect only the major labels and their affiliated international conglomerates. In the case of independents such as Third Man Records, Bartmanski and Woodward talk about the risks of raising "consumption anxiety" in buyers,

> a situation where the "sacred" activity involved in the independent production and release of music can become "polluted" by the visibility of profit-seeking and those who attempt to push up market prices. When profit-seeking behaviors are identified as being the prime motive behind vinyl production, the "sacredness" of the vinyl object is brought under suspicion (2015, 115).

Even as the object is mass-produced, it must provide for its users the aura of being an individual marker of his or her tastes and discrimination.

To be clear, however, as mass-produced objects, records have always walked this line between art and commerce. It is perhaps the tension between the poles of art and commerce that gives recorded music whatever vitality it might possess. Bartmanski and Woodward say, "[c]ommodities, such as the vinyl record, may be seen as 'sticky things' onto which all sorts of cultural and social meanings are fixed, and through which systems of embodied practices and collective activities are formed" (2015, 101). Perhaps, then, while still remaining the products of industrial processes, as containers for "art," records retain something outside of those processes, or as Sonja Windmüller says of Baudrillard, they "refuse to be part of the predominant function-based paradigm of interpretation" (2010, 49–50). Windmüller states of this refusal, "some groups of objects appear to be out of the ordinary in the utilitarian system, in concrete terms, 'unique, baroque, folkloric, and antique objects'" (2010, 50). Windmüller uses the term "cabinet of curiosities" to describe trash museums, places in which refuse is collected and displayed for aesthetic purposes, a term which is also used by Third Man to describe the wooden box in which the records of *The Rise and Fall of Paramount Records: Vol. 1 (1917–1932)* are housed (2010, 49). It could be that these commodities needed to fall into obsolescence in order to be recognized both as refuse of twentieth-century industrial capitalism and as objects of value worth preserving. Third Man Records not only recognizes the fetishism of the commodity that its elaborate deluxe packaging represents, but they seem also to celebrate it with all of its imperfections intact, in a process that might be called the fetishizing of fetishization.

One needn't search farther than the 2015 Record Store Day (RSD) to find an excellent example of how Third Man Records fetishizes fetishization itself. Record Store Day began in 2008 as a means of promoting independent music and independent record stores at a time when their very future was in question (Calamar and Gallo 2009, 26–27). As part of the promotion, record major and independent record labels release limited editions of special items designed to attract collectors into the stores. "Utilizing the actual unrestored audio as it currently exists on the sole original copy," one of the items Third Man offered on April 18 is a facsimile of Elvis Presley's first known recording, a ten-inch 78 of "My Happiness" b/w "That's When You Heartaches Begin" that Presley had paid $3.98 to record in 1953 (Third Man Records 2015). An acetate of the recording was auctioned off online by a high-school friend of Presley in January of 2015 for the highest bid of $300,000. The anonymous buyer was later revealed to be Jack White, and this acetate was the one Third Man Records faithfully duplicated down to the label and sleeve to offer collectors in April, or, as Third Man describes it on their website,

> From reproducing the typewritten labels (printed on the reverse of extra Prisonaires labels that happened to be laying around Sun Records in July

1953) to being packaged in a plain, nondescript, of-the-era sleeve, the utmost attention to detail has been paid in order to create an object so close to the historic original as to almost be indistinguishable from one other. (Third Man Records 2015)

In fact, one Facebook member of the YouTube Vinyl Community posted pictures and noted that, the "record here has all the scratchy pops and groove damage of the original which sourced this release. Also I have attached photos of what appears to be holographic damage to look authentic."[2] The photos show holographic "scratches" and other marks in the vinyl. Since only one copy of the Presley acetate is known to exist—indeed, Presley himself paid for only one copy to be made—the efforts of Third Man so closely to replicate that copy using advanced technology can be seen as perhaps an attempt to create a virtual aura in the mass-reproduced RSD release.

Why perpetuate such a hoax? What happens when our search for authenticity itself in the face of the cultural destruction wrought by the digital age can be packaged and sold back to us? And, if such is the case, how does it inflect the discussion of commodity fetishism? Have we finally reached Baudrillard's forth stage of simulation, in which the image "bears no relation to any reality whatever: it is its own pure simulacrum," in which even the pretense to reality can be falsified (Baudrillard 1988, 170)? Bartmanski and Woodward say of the kinds of small-batch production that draw the collectors on Record Store Day,

> In the terms proscribed by arch-postmodern theorist Jean Baudrillard, the vinyl in these markets is situated clearly beyond the logic of being a utility, that is, merely a medium for music listening, as there now exist a range of media for distributing music in massive numbers via digital files. Instead, rather than having utility value, it must be seen in terms of its sign value. (Bartmanski and Woodward 2015, 111)

In terms such as these, is it even fair to call what Third Man has done a "hoax," since the truth that the hoax would deny is impossible to access in the first place? In the case of an historical figure such as Elvis Presley, is it even possible to separate the man from the many myths that have grown around his legendary status, not the least of which is the formative myth of how he was "discovered" via the very record Third Man Records has carefully simulated?[3] A 2015 profile of Jack White in *Billboard* related to the Presley release says, "Third Man is part business, part cultural center and part artistic laboratory" (Levy 2015). About the vinyl resurgence that his company has come to represent, *Billboard* quotes White as saying, "You're reverential to it. With vinyl, you're on your knees. You're at the mercy of the needle. You watch the record spin and it's like you're sitting around a campfire. It's hypnotic" (Levy 2015). White is here calling attention to the sacral nature of the object, a point with which Bartmanski and Woodward agree. They note that in the digital age "vinyl becomes the epitome of 'warm' and 'humane' materialization of music, a perfect mode of making elusive music tactile, friendly, and more 'sacred' again" (Bartmanski and Woodward 2015, 32).

Calamar and Gallo describe the lure and the influence of the record store in terms that would not sound out of place when describing the vocation of a priest. They discuss "communities" of "fanatics" looking to their formative moment in the record store for "inspiration" (Calamar and Gallo 2009, 6–7). They quote the poet and singer Henry Rollins as saying, "[m]usic is such a shared experience. I believe in it as a great power for good. I don't believe in a higher power, but I do think that music is mankind's greatest achievement. Einstein was cool, but he's got nothing on Coltrane" (Calamar and Gallo 2009, 7). The question, though, is whether or not the sacral can be reproduced in another, perhaps "virtual," realm. Peter Bürger separates "sacral" and "courtly" art from "bourgeois" art, saying that unlike bourgeois art, the former two types of art are "integral to the life praxis of the recipient" (Bürger 1993, 238). In the bourgeois art of the modern industrial period, a "citizen who, in everyday life has been reduced to a partial function (means-ends activity) can be discovered in art as 'human being.' Here, one can unfold the abundance of one's talents, though with the proviso that this sphere remain strictly separate from the praxis of life" (Bürger 1993, 238–239). Praxis gives art a use value, specifically, the kind of value that Adorno bemoans the loss of under industrial capitalism. The complicated status of "My Happiness" as an object—in what exactly does its use-value lie?—tends to place it outside the bounds of any such definitions; if it is not sacral art in any traditional sense, is this merely the return of commodity fetishism in another register?

Even the way these reissue packages are compiled bespeaks the influence of the sacral as well as the museum, since they are often referred to as having been "curated" and comprising a "canon" of modern musical forms. Bartmanski and Woodward discuss specifically the "association of vinyl with canonical musical performances, which is "expressed in the current patterns of vinyl production, especially in the reissue of classic albums in rock, indie and jazz genres" (2015, 125). They single out the 4 Men With Beards label as offering "a finely curated catalog of reissues" (Bartmanski and Woodward 2015). A 2012 profile of the label notes that

> Most niche reissue labels focus on obscure artists who've retained rights to their music but never acquired proper distribution, which can simplify the reissuing process. 4 Men With Beards ventures into the murky bureaucracy of

licensing the rights to music that is controlled by major labels, but is tragically languishing out of print. (Lefebvre 2012)

While conceding that the lack of a focus on one particular genre of music robs 4 Men With Beards of a "distinct label identity," the profile of the label and its founder Filippo Salvadori says, "the careful curatorial selections will undoubtedly appeal to discerning listeners whose interests also shirk specific genre alignment" (Lefebvre 2012). The profile describes one such sacral object through which careful curation has resulted, the album *Second Edition* by Public Image Limited as, "a tremendous undertaking. Second Edition was originally released as three 12-inch records running at 45 RPM, contained in a metal film canister with P.I.L.'s distinct logo embossed upon the top, and 4 Men With Beards reissued it precisely that way" (Lefebvre 2012). This release was in keeping with the 4 Men With Beards policy of "faithfully mimicking the original artwork" of the albums they reissue (Lefebvre 2012).

The sound quality, though, is another matter, and 4 Men With Beards have been plagued with bad word-of-mouth about their quality control on online forums for years. One post on the Steve Hoffman Forum summarizes an earlier discussion about the quality problems of the label thusly,

To summarize what came up then:

1. the packaging is good, nice heavy sleeves, the vinyl is flat and heavy and looks good, but

2. it's pretty much certain that these are effectively CDs pressed to vinyl, after comparing a couple of their LPs to CD I couldn't hear the slightest difference and others have agreed with this. It's not an upgrade to buy one of their records.

3. their quality control leaves a lot to be desired, for example, recurring scratchy noises on Tim Buckley's Happy Sad

Conclusion: if you can find the music elsewhere, don't buy from them. (RHCD 2014)

Perhaps, though, to describe what you are doing in manufacturing reissues of industrial products in such a lofty way invites the desire to diminish that accomplishment? In a 2009 *New York Times* piece, Alex Williams talks about the increased use of the word "curate" to refer to activities and occupations having nothing to do with either museums or churches.[4] Williams notes that "[t]he word 'curate,' lofty and once rarely spoken outside exhibition corridors or British parishes, has become a fashionable code word among the aesthetically minded, who seem to paste it onto any activity that involves culling and selecting" (2009). Williams quotes the owner of a used clothing shop as saying that "invoking the word can be good for one's image and business" (2009). Karuna Tillman James, owner of "Curate Couture," says that the use of the word "curate" in the name of her business is a way of distinguishing her consignment shop from other second-hand clothing stores, saying that as opposed to "selling stuff that was gross and old and had been crammed in trunks for years," a curated shop "would have very specific pieces, selected purposefully" (Williams 2009). Bürger might be correct in asserting that in order to maintain its autonomy avant-garde art in the bourgeois era needed to divorce itself from praxis and everyday use, but that does not mean that praxis had to divorce itself from art. After all, only a very narrow definition of both "art" and "praxis" would disallow art in the practice of everyday life in the first place. We, in the West at least, live in a world of commodities that may or may not provide imperfect substitutions for our lost authentic existences, but it must also be acknowledged that those commodities are to some degree designed, and that human creativity need be employed in both the design and the marketing of these commodities that provides us with objects to fetishize.

In *The Practice of Everyday Life*, Michel de Certeau notes some of the ways that through consumption, praxis retains some of elements of creativity denied by industrial capitalism and commodity fetishism. According to de Certeau,

> [t]he imposed knowledge and symbolisms become objects manipulated by practitioners who have not produced them. The language produced by a certain social category has the power to extend its conquests into vast areas surrounding it, "deserts" where nothing equally articulated seems to exist, but in doing so it is caught in the trap of its assimilation by a jungle of procedures rendered invisible to the conqueror by the very victories he seems to have won. (De Certeau 2002, 32)

The last chapter discusses the ways in which collectors put this oppositional discourse to their own specific uses. The concern here is with the ways that those commodities walk the line between the knowledge and symbolisms they attempt to impose in the act of their consumption and the procedures by which it might be assimilated to other, perhaps oppositional ends. An example of such assimilation might be provided by a space operated by Bucks Burnett in Dallas, Texas, called The Eight Track Museum, a place referred to by Brad Farberman as "a charming slice of weirdness that descended on Dallas's Deep Ellum neighborhood in late 2010" (Farberman 2012). Farberman notes that while the Dallas location features "walls lined with eight-tracks, eight-track players, and eight-track advertisements," Burnett's collection does not limit itself to the eight-track tape. He quotes Burnett as saying that the museum also includes "the history of all recorded sound formats, from the 1800s when Thomas Edison invented the wax cylinder on up through the iPod" (Farberman 2012). While Burnett makes the point that his museum is as much about connecting with others over a shared love of music as it is about celebrating an outmoded form of technology, Farberman also calls the museum, "an ode to collecting, and the joys of discovery" (Farberman 2012).

Brian Durrans says of museum practices, "Any engagement with the world, in thought or in action, entails selecting what is relevant to the purpose in hand and rejecting what is not" (Durrans 1992). Durrans quotes T.O. Beidelman as saying, "all museum exhibitions are inherently problematical" (Durrans 1992). They are so, at least partially, because while they present a subjective interpretation of the "reality" they represent,

> In collections and displays, however, this irresistible force meets its immovable object. If everything else is fabricated by the enquirer, the bits of material, however classified or interpreted as ethnographic artefacts, are certainly not. Exhibiting or publishing such material amounts to a public assertion that a world exists independently of what we say about it. (Durrans 1992)

Timothy Luke says of exhibitions, "Museum exhibitions become culture-writing formations, using their acts and artifacts to create conventional understandings that are made manifest or left latent in any visitor's/viewer's personal encounters with the museum's normative performances" (Luke 2002, 3). Luke makes the point that by entering a museum exhibition, we learn something about ourselves in relation to the objects on exhibit. Because of this educational and sociological function museums, "operate as power plays in which plays for power circulate with the movement of viewers through their curated spaces" (Luke 2002, 3). He calls museums "modern scientific society's 'secular cathedrals,' 'guardians of shared history,' [and] 'storehouses for national treasures'" (Luke 2002, xiv). The social, ideological, and epistemological questions surrounding museums, however, are not limited to the spaces that identify themselves as such.

As this chapter noted at the outset, there is a curatorial function to the selection and the packaging of recorded music as well, and as such they are subject to the same sorts of critique as the museum proper. This is the essential tension enacted by the vinyl record as both an objective commodity and a performance subject to individual interpretation. As containers for musical performances, like vinyl records, eight-track tapes retain something of their commodity form, even when that form was itself consigned to the scrap heap of culture, but these obsolete objects still do provide access to the performances they contain, requiring only the technology with which to access them. Bartmanski and Woodward begin their study of the resurgence of the vinyl record by arguing that music exists more as an experience than as a thing, asserting that "[l]isteners always seem to bring an element of subjectivity with them," while "[t]he physical properties of sound 'objectively' shape what we hear" (2015, 1). As curated forms, perhaps deluxe reissues provide listeners with a ground on which to contest their ultimate significance, be it as a fetishized commodity, an authentic representation of cultural history and personal taste, or something in between.

Jonathan Sterne ties the rise of what he calls "audile technique" from which phonography emerges in the late nineteenth century to the "growing consumerist middle class" (2003, 160). Sterne calls audile technique "a bourgeois form of listening," and notes that it is

> [r]ooted in the practices of individuation: listeners could *own* their own acoustic spaces through owning the material component of a technique of producing that auditory space—the "medium" that stands in for a whole set of framed practices. The space of the auditory field became a form of private property, a space for the individual to inhabit alone. (2003, 160)

In this created auditory space, the recorded object brings the listener into terms of intimacy and ownership with the sound representations contained therein. When audiophiles talk about it, particularly in terms of stereo imaging, they call this created auditory space the "soundstage" of the recording, which John Atkinson defines in the following way,

> if you can record not only a sound but the direction in space it comes from, and can do so for every sound wave making up the soundstage, including all the reflected sound waves (the reverberation or "echoes"), then you will be able to reproduce a facsimile of the original soundstage, accurate in every detail. (Atkinson 2008)

The listener of a recording with a good soundstage should thus be able to "locate" the musicians in the imaginary space surrounding him or her rather than hearing the speaker(s) as the source of the music. While Atkinson argues that this phenomenon can only be produced in stereo, he qualifies that assertion in several significant ways. First of all, he makes the point that in all but the most ideal circumstances the soundstage can only ever be created partially and imperfectly. The second qualification has to do with conflicting philosophies of recorded music. One philosophy holds that in a recording the musicians, the room, and the ambience of the recording space should be reproduced as faithfully as possible. The second "is to treat the recording itself as the event, the performance, using live sounds purely as ingredients to be mixed and cooked" (Atkinson 2008). Even in such a case, "a wholly artificial, but nevertheless effective, soundstage hanging between and behind the speakers, which bears no relation to anything that might have existed in real life" can be created (Atkinson 2008). In theory, that is.

A paper published by The Anstendig Institute complicates the matter even further, noting that, "to call stereophony accurate sound reproduction is a falsification" (The Anstendig Institute 1982, 1984). The soundstage is in fact a subjective illusion created by the listener, because,

> if a system's reproduction of spatial relationships is accurate, one would have to know if the reproduction matches those relationships exactly as they were at the microphones during the recording. Since heads are differently shaped and no one can be in exactly the same place as the microphones, the spatial effects of direction, depth, etc., will be different for each person in the room. (The Anstendig Institute 1982, 1984)

In other words, the listener created subjective imaginary space provides the ground on which he or she projects the musical performances contained on the record. We consider this "space" to be our private property, as Sterne asserts, and it is precisely this "ownership" that makes recorded music such a slippery object for the recording industry to commodify. Baudrillard would say that this is precisely the point, that this simulacrum of acoustic space is the logical outcome of the triumph of hyperreality and the diffusion of power. He says that the predations of capitalism have so degraded anything having to do with reality that all that is left is "hysteria of production and reproduction of the real," which Baudrillard says, "seeks through production, and overproduction, is the restoration of the real which escapes it. That is why contemporary 'material' production is itself hyperreal" (1988, 180). If this is the case, it does not stop the record industry from asserting the true "thingness" of the objects it produces and places in the market, and perhaps to some degree it helps them.

Record Store Day has become a sort of "ground zero" for these battles over the meaning of the recorded object. What began as an attempt to draw customers back inside independent record stores continues growing every year in terms both of what they offer and the number of customers who show up and stand in line. The Record Store Day website, which went online at the same time as the first RSD describes it as follows:

> Record Store Day was conceived in 2007 at a gathering of independent record store owners and employees as a way to celebrate and spread the word about the unique culture surrounding nearly 1400 independently owned record stores in the US and thousands of similar stores internationally. There are Record Store Day participating stores on every continent except Antarctica. (Record Store Day 2015)

Billboard magazine reports that the 2015 Record Store Day was a "triumph," noting "the independent sector counted for 532,000 album scans, or 21.5 percent of total physical album sales, and 11.9 percent of overall sales. In both instances, it represented the highest weekly sales percentages total for the indie sector since 2003" (Christman 2015). Despite its stated intentions and perhaps because its success, Record Store Day is not without its naysayers, even among the vinyl aficionados to whom it was conceived to appeal. Bartmanski and Woodward make the not insignificant point that the meanings that we attached to vinyl records are at least partially inflected by the digital culture in which those meanings circulate, saying "one of the fundamental arguments of this book is that the widespread digitalization of music listening and music consumption has radically transformed the meaning of vinyl as a commodity" (2015, 102). The existence of online sites such as Discogs and eBay have complicated both the intentions of the retailers involved in RSD and the meanings attached to the record label wares on offer. Whether retailers or the record industry like it or not, Record Store Day has become an online phenomenon as much as it is a brick-and-mortar one.

Of one of the positive aspects of RSD, Emma Garland says that, "it introduces those who may not have considered buying a record before to an entirely different market of, what Stewart Lee likes to call 'flat, round mp3's'" (Garland 2014). Of course, this does not stop some manufacturers from releasing substandard products with bad quality control or other forms of capitalist greed. Garland is not alone in worrying that "the innate exclusivity of Record Store Day lends itself to exploitation" (2014). Peter Hinson quotes an open letter from Paul Weller to fans about why he will no longer be part of the promotion:

> This is a message to all the fans who couldn't get the new vinyl single on Record Store Day and/or paid a lot of money for a copy on eBay.
>
> I agree with all of you who have sent messages expressing your anger and disappointment at the exploitation of these "limited editions" by touts.
>
> Apart from making the record, the rest has very little to do with me but I am disheartened by the whole thing and unfortunately I won't be taking part in Record Store Day again.
>
> It's such a shame because as you know I am a big supporter of independent record stores but the greedy touts making a fast buck off genuine fans is disgusting and goes against the whole philosophy of RSD. (Hinson 2014)

The truly remarkable thing about this quote is that Record Store Day can still "dishearten" a veteran of almost forty years in the music business. A fundamental misunderstanding—whether willful or not—of the nature of market capitalism seems to guide such reactions to the phenomenon of RSD flipping, or of what Weller calls "touts" getting in at the front of the line, snapping up the exclusive, limited offerings early in the day, and selling them later on eBay at an exorbitant profit. Surely this misunderstanding must be the work of what Lukács meant when he defined the workings of reification as "a relation between people takes on the character of a thing and thus acquires a 'phantom objectivity,' an autonomy that seems so strictly rational and all-embracing as to conceal every trace of its fundamental nature: the relation between people" (1971, 83). Surely the "phantom objectivity" to which Lukács refers lies behind the assumption that commodities might defy their very nature in the name of a perhaps more altruistic motive?

What prompts the following reaction to Record Store Day from someone presumably as wise to the ways of capitalism and exploitation as is Henry Rollins? Bartmanski and Woodward quote Rollins speaking about what they call "the fetishization of the collector's impulse" as follows:

> Limited edition, colored vinyl, 7-inch single with a non-LP B-side? Wait, it gets better. The first of three singles all featuring the same A-side but different B-sides. Hold on—the 12-inch version comes with a live track and a demo version of the A-side, but not the non-LP B-side that's on the 7-inch. To hear it all, you have to get all six releases. If you are someone burdened by real life, all of this is boring and yet another example of the cruel and unusual machinations of predatory capitalism. (2015, 115–116)

Perhaps Rollins is imagining a capitalism that is not cruel and unusual in its machinations nor as predatory? But, is it not in the very nature of capitalism to be predatory? To put profits before human relations? Lukács says that "the individual object which man confronts directly, either as producer or consumer, is distorted in its objectivity by its commodity character," but that "the reified mind has come to regard them as the true representatives of his societal existence" (1971, 93). It would be truly remarkable, in fact, if record companies and scalpers did *not* attempt to exploit the desires and dreams of collectors on a day dedicated to valorizing the retail impulse. Our flaw in this sense is misrecoginzing capitalism in its true clothes, is it not?

Or is it? Is it possible that in the case of recorded music, there might be some excess of energy that escapes the clutches of consumer capitalism? Is there, in fact, a non-fetishistic way to relate to these objects that does not commodify them? Is there a use-value that defies the logic of exchange and rationalization inscribed in the very nature of the form? Sam Lefebvre, who writes an occasional blog for *San Francisco Weekly* called "Record Peddler," says of Record Store Day 2013, "[l]ater in the day at 1-2-3-4 Go!, many earnest patrons spent money, hung out, ate from food trucks out front, and the event resembled what it's intended for: a reason to get citizens in local record shops" (Lefebvre 2013). However, if the intent is not to get those citizens to spend money, then what is it? Lefebvre points to the difference when he says, "Record Store Day urges fans to patronize these stores, at least for a day, if they wish to buy special titles from their favorite artists" (2013).

The difference appears to be one of values. "Fans" value the objects the record labels release on this day in a difference register from the flippers and speculators of whom Lefebvre notices on the morning of RSD 2013 that they were already selling the specially released boxed set of Dave Matthews Band live recordings that listed at $79.99 retail for $500 online (2013). The "fans" Lefebvre speaks of could be the ones identified by Bartmanski and Woodward as those for whom "vinyl is an important symbol of expertise, distinction and skill" that extends from those who make music as DJs and professional musicians to the collectors who keep the music alive (2015, 117). They quote Keith Richards as saying that making records was a way of legitimizing musicians who had professional aspirations, and that holding the object became a symbol of that process (Bartmanski and Woodward 2015).

Records not only provided the stamp of legitimacy to a performer, but they also made the dissemination of many types of music possible to musicians in the first place. The Beatles appearances on US television may have sparked the imaginations of teenagers all over the country that grew their hair and started bands, but it was the Beatles records that taught them how to play. Mark Katz makes a similar point about jazz when he says, "[p]honographs and records, small and easily transportable, gave budding musicians unprecedented access to jazz. Without this feature of recording technology, some jazz artists might never have pursued their careers" (2004, 74). Non-musicians also use the objects to provide legitimacy to their pursuit of the knowledge and the understanding of musical forms. Knowing the "best" pressing of a particular recording is in itself a particular form of competence, whether or not one considers it to be the equivalent of playing the violin or the electric guitar. Bartmanski and Woodward point to the objects lending "authenticity" to the pursuit of collecting, and they note that it unites musicians and non-musicians. They say, "vinyl becomes a symbolic commodity, possession or production of which points to certain knowledges and expertise" (Bartmanski and Woodward 2015, 118). Vinyl records, in fact, are among the unique cultural objects in which symbols assume actual physical form. Holding the record up to light presents the viewer with the symbolic representation of the musical "performance" contained therein inscribed into its very grooves.

At the same time, it cannot be argued that records are not also mass-produced commodities that are subject to the same conditions of capitalist exploitation and alienation as any other commodity. If, on the other hand, it *can* be argued that the culture represented by the independent record store can be said to transcend at some level the violence that capitalism does to cultural forms and human relations, it must be on the level of the performances contained within the commodities that these record stores offer consumers. Despite the rise of digital downloads and streaming services and other attempts to break it, however, that connection of performance and object has proven to be remarkably robust in the imaginations of musicians, DJs, and record collectors the world over. Record collectors, in particular, have come to represent a distinct identity at a time in which the mainstream identification with music is perhaps becoming more tenuous. In an ironic twist, Katz worries that "the intangibility of MP3s and the ease with which they are obtained, disseminated, and deleted may encourage the sense that music is just another disposable commodity" (2004, 175). Recorded music will have had to be something other than that in order for such a danger to exist, though.

Vinyl records as objects carry enough of a cultural charge beyond being commodities that even in the digital age online communities are growing around them, the subject of chapter 7 of this book. For now it can be said that they help police the line between use-value and exchange-value in vinyl records. Just before Record Store Day 2015, several vinyl special interest groups on Facebook were organizing to find ways to beat the flippers, their members offering to pick up available scarce reissues and sell them to each other at cost. One such group, called Record Collector's Paradise includes in its bylaws the rule that "RSD product will only be accepted for sale on this site provided it is offered at its original cost plus local taxes for the first 6 months after the RSD it was released on. After 6 months, it can be offered without restriction" (Campbell 2015). As for the reasons such resistance exists in the first place, Calamar and Gallo quote John Kunz of Waterloo Records in Austin as saying,

> we have these wonderful analog listening devices on the sides of our heads that don't want to hear zeroes and ones. They want sound waves, a human-sixe arc. Vinyl is always going to have a place, especially among the people who work in this store. There is a tremendous romance for the record. (2009, 187)

They also quote Susanna Hoffs of the Bangles saying "I always liked the closeness of listening in my room to my turntable, getting the whole statement of the artist. I can see why kids are gravitating toward that" (2009, 187). Furthermore, "the day is a reminder that the record store is a hangout, a community center, a place to hang out with like-minded people" (Calamar and Gallo 2009, 224).

That ability to capture individual imaginations and build communities sets the vinyl record apart from being a mere fetishized commodity, even as it maintains its commodity form and features. Hoffs's point about being in her room listening could be extended to encompass the "personal acoustic space" to which Sterne refers, an individual space that all listeners carry around inside their heads. Analog discs map this space out for listeners in the grooves etched within. As captured "performances," vinyl records retain the ability to also capture the imaginations of listeners and collectors. Bartmanski and Woodward call this fetishization of another valence, calling it a "passion [that] stems partly from a human need, or even carnal desire, to ground our sense of reality through attachment to actual, concrete objects made by other humans, for example, manufactured commodities" (2015, 133). People who associate themselves with vinyl records at the level of collectors construct an identity around these objects both as commodities and as containers for musical performances, which is the subject to which we will turn in the next chapter.

NOTES

1. It is absolutely clear that, by his activity, man changes the forms of the materials of nature in such a way as to make them useful to him. The form of wood, for instance, is altered if a table is made out of it. Nevertheless the table continues to be wood, an ordinary sensuous thing. But as soon as it emerges as a commodity, it changes into a thing which transcends sensuousness. It not only stands with its feet on the ground, but, in relation to all other commodities, it stands on its head, and evolves out of its wooden brain grotesque ideas, far more wonderful than if it were to begin dancing of its own free will. (Marx 1977, 163)

2. John Beaulieu's Facebook Page, accessed April 18, 2015, www.facebook.com/megatrendsinbrutality

3. In 1953, he appeared with his beat-up guitar in the office of Sun Studios in Memphis. He wanted to make a record to surprise his mother, he said, but "if you know anyone that wants a singer ..." "Who do you sound like?" the receptionist asked. "I don't sound like nobody." Elvis recorded two Ink Spots songs, "My Happiness" and "That's When Your Heartaches Begin," with Sam Phillips, the owner of Sun, listening from the control room (Altschuler 2003,25).

4. Michael James Vechinski also discusses the Williams article in his chapter from *Contemporary Collecting* titled "Collecting, Curating, and the Magic Circle of Ownership in a Post-Material Culture" (Vechinski 2013).

Chapter Six

"You Spin Me Round (Like a Record)"

Analog Audiophilia as Disciplinary Mechanism

The first episode of the second season of the television show *Lost* featured the introduction of the Dharma Initiative and the character Desmond, who spent his days in the hatch, pressing a sequence of numbers into a computer.[1] The episode opens with Desmond awaking to the beeping of the computer, signaling that 108 minutes have passed and that he must type the numbers into the computer. A montage follows of the character, who is unidentified at this point, performing his morning routine, which includes cleaning up, exercising, bathing, and dressing. Aside from the fact that he has to input the numbers into the computer, two other things make Desmond's routine worth remark. First, he appears to inject himself with a drug contained in a bottle with the same numbers on the label that he inputs into the computer. The second remarkable thing Desmond does after inputting the numbers is to put a vinyl record on the turntable and play Cass Eliot's song "Make Your Own Kind of Music." The entire sequence makes it clear that this is Desmond's routine in the automatic ways in which the character moves around in what we eventually come to know is a research station called The Swan beneath the hatch that had so concerned the castaways throughout the first season of the show. Viewers later in the second season find out that pressing the numbers into the computer every 108 minutes prevents the release of an electromagnetic pulse and that the numbers hadn't been pressed during Desmond's death struggle with his predecessor forty-four days before, at which point the resulting pulse brought down Oceanic Flight 815 and brought the survivors to the island.

This scene resonates both within the narrative of the episode and of the series. What is it meant to represent? First, it introduces the character Desmond, who will play a significant role in the saga as it unfolds, specifically an ethical role. Second, the scene is meant represent the disruption of the routine that Desmond has set up for himself, a disruption with potentially deadly consequences for both Desmond and the survivors of Flight 815. Third, while the musical accompaniment to the scene provided by the record Desmond plays comments on both the scene and the overall narrative of *Lost*, the fact that the character knows how to operate the turntable properly is itself a comment on how he has integrated his identity into the routine of the Swan Station, suggesting ways in which disciplining the actions of a listening subject to the requirements of listening to vinyl records play into the process of enculturation and the formation of identity. Vinyl and shellac records are fragile commodities that require a kind of specialized knowledge and care that other subsequent music formats either do not require or require to a much lesser degree. Collecting and caring for vinyl records, therefore, constitutes a kind of discipline in the Foucauldian sense requiring the type of governance Mitchell Dean refers to when he refers to "conduct" as, "any attempt to shape with some degree of deliberation aspects of our behavior according to particular sets of norms and for a variety of ends" (1999, 10). To some degree, the inclusion of the turntable and the age of the record are intended to contribute to the uncanny feeling of the décor of the Hatch, wherein objects from decades dating between the late 1960s and the mid-2000s are represented. The deliberately archaic inclusion of the turntable makes the character's understanding of how to use it—since it was widely believed in 2005 to be an obsolete technology—all the more curious.

The automatic physical actions that make up a daily routine, this chapter will argue, play a not insignificant role in enculturation and identity formation. Via Foucault, Bratich, Packer, and McCarthy refer to this process as "governmentality," or "the conduct of conduct," which "takes place at innumerable sites, through an array of techniques and programs that are usually defined as cultural" (2003, 4–5). Tony Bennett defines culture as,

> a historically specific set of institutionally embedded relations of government in which the forms of thought and conduct of extended populations are targeted for transformation—in part via the extension through the social body of the forms, techniques, and regiments of aesthetic and intellectual culture. (Bennett 1992)

As this scene makes clear, the bodily movements and the knowledges they contain that contribute to the makeup of a routine become embodied techniques, sites in which muscle memory both displays and disguises the relations of power and knowledge that they contain. In the very movements of his body through the space of the Hatch, Desmond enacts a worldview, a worldview that subsequent narrative events and revelations will show to be only partially "true." Desmond believes that if he does not input the numbers into the computer, something bad will happen, and events prove that fear to be well founded. However, he also believes that the air outside the Hatch is contaminated, and that he must inject himself from the vile with the numbers and wear a hazard suit when he goes outside, neither of which turn out to be completely true.[2] True or false, the ways that Desmond acts and moves through the Hatch—the ways in which he conducts himself—at the start of this episode make up his world as he knows it, including the record and the turntable, and through those actions, power both acts on and through him. As a technology, the turntable itself is just about as complex a mechanism as any Desmond handles in this sequence, and it requires special fields of knowledge and ways of acting in order to operate it properly. Adopting the technology requires physical learning as well as a mindset, and these ways of thinking and acting together help to form the "listening subject."

The lyrics to the Godley and Creme song "I pity Inanimate Objects" contain the lines, "The fewer the moving parts / The less there is to go wrong." Although these sentiments are expressed musically by means of a vinyl record revolving on a turntable, they assuredly do not describe the technology that reproduces them.[3] The record and the turntable on which it is played are both complex technologies that require special ways of handling and maintaining in order to function effectively. Thus there have arisen discourses involving the "right" and "wrong" ways of handling them. Mitchell Dean calls the concerns of such discourses "the ethical government of the self," and he defines four aspects of such government:

> First, it involves ontology, concerned with *what* we seek to act upon, the *governed or ethical substance* . . . Second, it involves ascetics, concerned with *how* we govern this substance, the *governing or ethical work* . . . Third it involves deontology, concerned with *who* we are when we are governed in such a manner, our "mode of subjectification," or the *governable or ethical subject* . . . Fourth, it entails a teleology, concerned with *why* we govern or are governed, the ends or goal sought, what we hope to become or the world we hope to create, that which might be called the *telos of governmental or ethical practices*. (1999, 17)

In the case of vinyl record aficionados, the questions of why, how, who, and what are all tied to the apparatuses and practices centering on the maintenance and care of the vinyl record and the device on which it is played. Dean makes the point that all of these practices "presuppose some goal or end to be achieved," and he also makes the more significant point to this discussion that such practices both base themselves on "truths about our existence and nature as human beings," but that on the other hand "the ways in which we govern and conduct ourselves give rise to different ways of producing truth"

(1999, 18). The "truth" that is produced in this case is the much-contested truth about the best ways to store, play, and care for vinyl records.

The production of truth requires specialized forms of knowledge and understanding of how to act upon them. In a discussion of the differences in which Stuart Hall and Nikolas Rose conceive of culture, Bratich et al. note

> It is through the deployment of particular forms of expertise in particular relations of government that particular ways of speaking the truth and making it practical are connected to particular ways of acting on persons—and of inducing them to act upon themselves—which, in their turn, form particular ways of acting on the social. (Bratich, Packer, and McCarthy 2003, 54)

The ways of speaking the truth about the objects and their uses in the case of vinyl records, and the particular ways of inducing people to act upon themselves in the use and maintenance of these objects and the devices on which they are played are the subject of this chapter. Foucault himself talks about what he is doing in terms of "eventalization," which he says, "means rediscovering the connections, encounters, supports, blockages, plays of forces, strategies and so on which at a given moment establish what subsequently counts as being self-evident, universal and necessary" (Foucault 1991, 76). Consider this, then an examination of the eventalization of discursive practices surrounding the use and care of records, shellac and vinyl. Consider also, that while all records require specialized knowledge in how to care for and handle them, the closer such discursive practices get to the realm of audiophilia, the more exacting and contentious are the requirements for the conduct of listening subjects.

The act of listening in itself, inasmuch as it partakes of what Jonathan Sterne refers to as "audile technique" can be seen as disciplinary in this sense. Sterne argues that techniques of listening orient themselves around "what sociologists and anthropologists have come to call '*the habitus*,'" which he further defines as "socially conditioned subjectivity" (2003, 92). Habitus involves "a mix of custom, bodily technique, social outlook, style, and orientation" (Sterne 2003, 92). Chapter 1 discussed the ways in which the discourse of fidelity embodied in the "Edison Realism Test" subjected listeners to a self-policing mechanism, through early listeners of recorded sound might have oriented themselves to the recording apparatus. It also noted that listeners can be taught to recognize high fidelity, and that classes actually exist in order to teach such listening practices (Kornelis 2015). Sterne notes that at the beginning of the history of technologies of sound reproduction, "[p]eople had to learn how to understand the relations between sounds made by people and sounds made by machines" (2003, 216). In discussing the training of early telephone users, he makes a point that is worth extending to the entire history of listening techniques as they relate to recording technology, saying that listeners, "were to hone their audile technique, to become connoisseurs of the various shades of perfection in tone, thereby learning to distinguish between truth and falsity, or at least to be able to construct their own auditory realities" (Sterne 2003, 267). Shellac and vinyl records require just such listening practices, and thus help to construct the auditory realities of listening subjects.

Evan Eisenberg illustrates this point when he asks his Uncle Saul whether or not there were moments in which his relentless chase for sound perfection justified his efforts. Saul responds,

> Moments. Moments. But they were transitory. Because just at the moment when you felt secure, another external force said to you, "Uh uh. Not yet . . ." But when it was good, it was very, very good. When I put on a test record on and it said "Thirty-five cycles," and my speakers went—he paused, listening for the tremor—and when it said "Twelve thousand cycles" and my ears started to ring—I knew it was there, you know what I mean? I had: No hum. No surface noise. No distortion. With the controls and all of the variables that I could bring to bear on it, I could filter out or add or subtract, fine-tune each record to take care of its particular deficiencies. (Eisenberg 2005, 185)

If Eisenberg's Uncle Saul comes to the conclusion that the transitory moments he was chasing were ultimately quixotic, it might be the nature of the medium through which he was chasing them that is to blame (Eisenberg 2005, 184). Sterne makes the point that a "medium of sound reproduction is an apparatus, a network" involving the relations between many people, practices, and technologies (Sterne 2003, 225). Saul uses his expensive equipment to negotiate a record's "particular deficiencies," but in order to find something deficient one needs a "whole" with which to compare it.

Chapter 2 discussed Benjamin's concept of "aura" in mechanical reproduction, along with its concurrent notion that the closer one gets to the "original" the less that said aura is debased. Jonathan Sterne cites Benjamin to argue that the "possibility of sound reproduction reorients the practices of sound reproduction; insofar as it is a possibility at all, reproduction precedes originality" (2003, 221). Thus, the whole conception of an "original" of which the recorded version is a "copy" is itself a product of the medium. In addition to teaching the listening subject what he or she was listening for, the audile technique engendered by recorded music also needed to teach listening subjects how to listen for it, what constituted the recorded sound through which listeners were sorting to find the "original;" "[i]n developing their audile technique, listeners learned to differentiate between sounds 'of' and sounds 'by' the network, casting the former as 'exterior' and the latter as 'interior' to the process of reproduction" (Sterne 2003, 283). What was true at the start, when the limits of the technology made a necessity of practices of disciplining the ear is still true today, perhaps uniquely true of vinyl recordings, since surface noise of one or another sort is an automatic byproduct of the physical connection between the stylus and the groove.

Chapter 1 referenced André Bazin's "myth of total cinema," which holds that film is working toward equivalence with actual experience, a development that would mean its own erasure as a medium. Sterne makes the same point about fidelity, a desire he says "would lead to a conflicted aesthetic of reproduced sound, where the ideal state for the technology as vanishing mediator would continually be set in conflict with the reality that sound-reproduction technologies had their own sonic character" (Sterne 2003, 225). Setting aside for the moment the idea that such a desire could never realistically find fulfillment, selecting, setting up, and balancing the equipment this ideal would require, adjusting it to the sound properties of the space and finding the "sweet spot" at which all of the expense and effort might meet its mark, would the listening subject have to efface all that had brought him or her to that ideal state in order for the mediator to "vanish." The same audiophiles who may desire the elision of the medium through which such idealized sound is elided seem to have little problem fetishizing the expensive equipment, apparatuses, and techniques to reach that state. Furthermore, some listeners actually prefer vinyl records technology because of its imperfections. Yochim and Biddinger quote a young record collector as saying

> I like the popping and the crackling to an extent. Some of them are so bad that you can't really hear [the music], but if it's in moderation, that little fuzz that you get from a record, I like that.... If you're listening to an old record, it'll kind of have an old sound to it. So if you're listening to a record from 1960, you have a record that sounds like it's from 1960 [*sic*]. In a good way, though. It kind of gives you that vintage feel. (Yochim and Biddinger 2008)

This can hardly be said to be the voice of someone who desires the medium to disappear. The fact that vinyl takes so much interaction and physical activity to play and maintain mitigates against any desire to elide the technology in its listeners; it is difficult to erase the medium for long when the listener must get up in order to change the record every two to twenty minutes or so.

What can be seen as remarkable is the return to this imperfect medium in the face of technologies such as digital and streaming music that bring listening subjects so much closer to being able to efface the medium entirely, technologies that require relatively little disciplinary effort to operate, the main worry about streaming devices seems to be whether or not the battery is charged. Meanwhile, discipline is part of the appeal of vinyl technology, and it has been built into the very nature of vinyl from the start, since both the records and the equipment on which they are played are relatively fragile and easy to damage, a peril that grows the more sophisticated and expensive the technology becomes. Vinyl aficionados choose the technology that suits them, and they construct their identities as listeners around the physical limitations and possibilities of the medium. Most of the listeners asked for this study about the care of their records and equipment as a factor in their adoption of vinyl technologies both identified such care and effort as important to them and also to the appeal of vinyl in the first place. One listener responded when asked about how he cares for his records, "Regular required cleaning and upgrades such as stylus. I have a Nitty Gritty record cleaner and i mix my own solution with alcohol and distilled water. I handle everything with utmost care and make sure everything is stored in proper sleeves" (S. Del Siblock, personal communication, December 29, 2015). Another listener responded, "I clean most (especially used) records on a VPI 16.5 record cleaning machine, then put the cleaned record into an anti-static inner sleeve. Once in its jacket, that goes into a 3mil outer sleeves" (R. Clark, personal communication, December 29, 2015). A third respondent said that he uses a "Decca Record Brush, [an] Anti-Static Brush from Mapleshade, [a] Styli brush, [and] sometimes Discwasher" (M. LaScola, personal communication, January 2, 2016). These responses might be considered somewhat representative of the efforts to which some vinyl listeners will go in order to optimize their listening practices, and of the discipline such practices require. What should be clear at this point is that although there is not one "method" used to care for one's vinyl treasures, each individual has a method that he or she uses, and having a method is considered by many to be an important aspect of owning and listening to vinyl records.

Whatever might have been the original intent of the creators of these analog technologies, vinyl enthusiasts to some degree create the technologies and the techniques as they establish their own identities as such. Bartmanski and Woodward call attention to the physical presence of vinyl, saying that "[i]t invites one to ritualize and celebrate the act of listening" (2015, 58). Yochim and Biddinger also draw the contrast directly:

> Vinyl collectors talk about the actual qualities of records: the surface noise, their mass, the artwork on the covers. These qualities were always present in vinyl records, but until vinyl's dominance was challenged by tapes and CDs, records could not be seen for their connection to people's humanity, their fallibility, and their mortality. (Yochim and Biddinger 2008)

They quote another collector as saying, "The moving parts of a turntable and the ability to see what's going on allow you to participate more in the music" a participation that is more or less required by the medium's delicacy, about which the collector continues, "You're forced to be careful; you're forced to get the record out and wipe it off and put it on" (Yochim and Biddinger 2008). These ritualized behaviors become part of the identity of the vinyl listening subject. Bartmanski and Woodward call vinyl a "high sensitivity medium," noting that "[v]inyl is sensitive to human (mis)treatment and to the effects of technical mediations, and it sensitizes us in turn to the qualities of these effects" (2015, 59). The medium not only refuses to disappear in this sense, but rather it forces itself into the listener's subjectivity from the very start.

There are many points in the chain of sound reproduction that require the intervention of a knowledgeable listener, the violation of which can damage either the hardware or the software, interfering with the sound or even rendering it inoperable. Vinyl and shellac records require a sharp object, the stylus, to come into contact with the grooves of the record. The grooves are made by "a needle vibrating on a cutting lathe carved that groove in a way directly related to the sound vibrations originally created from the instruments and voice boxes of the musicians who recorded the music" (Waehner n.d.). Because this point of contact is so important to the quality of the sound the listener receives from an analog recording, special care must be taken to preserve its integrity. Therefore, care must be taken to maintain both the record and the stylus so as not to interfere with the quality of the contact they make. The specialized knowledge and activity required to maintain and play records can be said to be a discipline, if not in the academic sense, then in the general, more Foucauldian sense of training "the moving, confused, useless multitudes into a multiplicity of individual elements—small, separate cells, organic autonomies, genetic identities, combinatory segments" (Foucault 1979, 170). With the "correct" training, listening subjects can learn the proper uses of their records and the apparatuses on which they are played, training that was originally provided by the manufacturers of the records and the machines but eventually came to be a more or less self-policing subculture in itself.

The instructions begin, "Now that you own a Victrola, a whole world of music is open to you" (Victor Talking Machine Co. 1924). The "world of music" to which the instructions refer is that offered by Victor Records, and the instructions make a peculiar promise to listeners, that "[t]here is no variety of personal taste and no condition of mind, to which the Victor records will not minister" (Victor Talking Machine Co. 1924). Purchasers of a Spring Motor Type Victrola in 1924 would also have been presented with rather a daunting set of instructions, even if they were phrased in a friendly and efficient way. The instructions and the machine are complex, even though the instructions themselves state that since the machine is "carefully constructed, and before leaving the factory it is subjected to the most rigorous tests," "[t]he instrument itself will not require any great or expert care" (Victor Talking Machine Co. 1924). The instructions proceed to introduce the parts of the Victrola and to present ten steps for setting up the machine out of the box and two steps for adjusting the pneumatic cushion lid.

The potential trouble spots that needed special care according to the instructions included getting the turntable to operate at the proper speed, which required some adjustment with a screwdriver; the mica diaphragm that vibrated in sympathy with the needle as it traced the groove, producing the sound that was then amplified by the sound box, which could be damaged by poking it with a finger or a sharp object; and the needle itself, which the instructions recommend changing with every play if it is a steel one in order not to damage the records. The only maintenance the instructions recommend is infrequent oiling of the motor and the mechanical parts. The instructions also include a brief section about caring for records, which recommends not stacking them and keeping them clean and free from dust. The act of learning to set up and use the Victrola can be seen as a means of training in the proper discipline of using the technology, and as Foucault says, "[f]or the disciplined man, as for the true believer, no detail is unimportant, but not so much for the meaning that it conceals within it as for the hold it provides for the power that wishes to seize it" (1979, 140). This set of instructions, and the thoughts and actions it engenders, can thus be seen as discursive in the Foucauldian sense, whose "practices are not purely and simply ways of producing discourse. They are embodied in technical processes, in institutions, in patterns of general behavior, in forms for transmission and diffusion, and in pedagogical forms which, at once, impose and maintain them" (Foucault 1977, 200). Foucault makes the further point that such practices are transformative, leading to changes in "mentality, collective attitudes, or a state of mind" (Foucault 1977, 200). The result of these instructions is an economy of power, control over oneself and oneself in relation to the technology.

Chapter 1 discussed the need for new technologies to be "tamed" in order to be used in the home. Part of taming the technology is learning how to use it and to integrate it into the family and the home. Knut H. Sørensen discusses the approach of industrial sociology in relation to the analysis of technology and human interaction in which "one was interested in the interaction of humans and machines related to processes of mechanization and automation. A machine may be seen as an arrangement that requires certain tasks to be performed" (Sørensen 2006). He argues that seeing this relationship only in terms of the impacts that technologies have on society is partial at best. Instead, he suggests via Bruno Latour and actor-network theory that the force of technology "emerges from the way technology and culture become enmeshed through delegation and re-delegation of action between human and non-human actants," concluding that "[w]e experience the force of technology through learning and discipline made invisible" (Sørensen 2006). Sterne makes the point that early on at least, "sound reproduction technologies worked only with a little human help" (2003, 223). Sterne is referring specifically to the efforts both of those demonstrating the new technologies and the auditors listening to those demonstrations, presenters using "heavily conventionalized language" that would be easier to make out above the other noises made by the machines, and listeners making the extra effort required to sort out the "sounds" from the "noise" of the recording and playback apparatuses.

Early versions of these technologies required physical and mental effort in addition in order to get the machines to work to the point that the sounds could be sorted out. The operating instructions for a Berliner Gramophone circa 1895, which opened with a photograph of a young girl using the machine, included the following admonition: "The American Hand Gramophone reproducer is a talking machine which is both simple and effective, and will not get easily out of order, provided that the following directions are carefully kept in mind" (Berliner's Gramophone 1894).[4] Even winding the mechanism that caused the turntable to spin required practice to get to "a wrist movement at the rate of about 150 times a minute," according to these instructions, which recommend "To acquire this regularity of motion, practice it a number of times with the lever and sound box lifted from the turntable" (Berliner's Gramophone 1894). Furthermore, one needed to "[h]old the handle loosely, so that it slides readily through the fingers" in order to practice these specialized movements. This manual also gives users instructions on how to grind a worn needle, and suggests the possibility of using a "'Thorpes' No. 14" darning needle as a replacement (Berliner's Gramophone 1894)!

Such attention to detail in this set of early gramophone instructions places them well within the purview of industrial capitalism, or of what Foucault might call the emergence of modern disciplinary society. Foucault says that "[d]iscipline defines each of the relations that the body must have with the object that it manipulates. Between them it outlines a meticulous meshing" (1979, 152–153). Through the minute movements instructed, a microphysics of power emerges in the interactions between human and machine, and in the case of the contact between stylus and groove of a sound recording, microphysics shrink down to the microscopic level. He talks about these interactions between body and object—"hold the handle loosely"—using the example of the military "*manoeuvre*," which he defines as "the instrumental coding of the body" (Foucault 1979, 153). Foucault uses the example of teaching a soldier how to handle a weapon, saying

> It consists of a breakdown of the total gesture into two parallel series: that of the parts of the body to be used (right hand, left hand, knee, eye, elbow, etc.) and that of the parts of the object to be manipulated (barrel, notch, hammer, screw, etc.); then the two sets of parts are correlated together according to a number of simple gestures (rest, bend); lastly, it fixes the canonical succession in which each of these correlations occupies a particular place. (Foucault 1979, 153)

Foucault notes that the power that emerges works through the bodily interactions of the subject with the machine, what he calls a "body-weapon, body-tool, body-machine complex" (1979, 153). He calls this power a function of "synthesis," saying that in place of exploitation, it provides "a coercive link with the apparatus of production" (1979, 153). The subject is not only practicing movements and their interactions with objects and machines, but he or she is also practicing a worldview. Foucault calls this the "natural body," and he describes it as "the bearer of forces and the seat of duration; it is the body susceptible to specified operations, which have their order, their stages, their internal conditions, their constituent elements" (Foucault 1979, 155).

Prior to the introduction of a governor to control the speed, the Berliner gramophone took a lot of practice in turning the handle in order for it to sound right, since variation in speed affected the tone and pitch of the recording. Even after the introduction of the governor, the machine still required practiced movements in order to function properly and not damage the machine or the recordings. The Berliner instructions orient the bodies of listening subjects in relation to the machine not only in the practice that it would take to operate the hand wheel and the velocity of the turntable properly, but also how to orient the machine in relation to the room and the body in relation to the machine. Number ten on the numbered list of instructions notes that "[h]angings and carpets deaden the sound from the horn; turning the latter close against the wall or door or wooden partition or against the corner of a room will heighten the effect and in particular will enable *the person turning the machine* to hear it well" (Berliner's Gramophone 1894, emphasis in original). This concern for the position of the machine, the room, and the listening subject in relation to each other continues to the present day. In fact, it became an even greater concern as the apparatus becomes more complex and after the introduction of stereophonic recording and playback in the late 1950s.

Sterne makes the significant point that buying into the discourse of fidelity is to a great degree also buying into the system of network of sound reproduction and the medium through which it is reproduced. The point at which these discursive practices become disciplinary is the point at which the listening subject begins policing the room, the devices, and the ear for "accurate" sound reproduction. The booklet *Understanding High Fidelity* that was discussed in chapter 1 suggests that by 1953, "high-fidelity art has progressed to the point where the electrical interconnection of various components in a system can be mastered by anyone who can plug a lamp into a wall socket" (Biancholli and Bogen 1953, 43). They include an extensive note accompanying a figure illustrating optimal speaker placement in the listening room that begins,

> We have to separate low frequency reproduction from high frequency in considering [speaker performance] because each imposes its own requirements, but obviously the best location is one that satisfies the needs of both. For low notes the best results come from placing the unit in the corner of a room. The next best choice is at floor (or ceiling) level, and, as can be seen [in the figure], the worst result comes from placing the unit in the middle of the wall suspended between the floor and ceiling. . . . High notes, because they are directional, should not be blocked off from the listener. Upholstered furniture in particular will absorb a good deal of the high-frequency energy radiated by the loudspeaker, so the arrangement of the room should not be such that furniture obtrudes between your ear and the loudspeaker. (Biancholli and Bogen 1953, 45)

Early in the stereo era, much ink was spilled on educating listening subjects about the sound reproduction apparatus. Tim J. Anderson makes the point about early high-fidelity listeners that

> the postwar, high-fidelity public was equipped with better playback equipment, a greater ability to participate in consumer culture, and an interest in the effects of scientific research at the popular level. In some ways the call for music and sounds that were "realer than real" was a twofold negotiation between record producers and audiences. (2006, 112)

Writing about loudspeakers in 1961, Edgar Villchur says they are "not necessarily less 'transparent' to the original sound than are all pickups, tone arms, and amplifiers, but in a typical high-fidelity system it is the loudspeaker that is most often the weakest link" (1961). He notes that in comparing the sounds of various audio devices "there is only one standard, that of *accuracy*" (Villchur 1961, 60, emphasis mine). Villchur says that the human ear is "the most valuable single tool among audio-measuring instruments," and he suggests buying a concert ticket before experimenting with selecting and setting up loudspeakers in order to establish a standard by which to judge the accuracy of the reproduction that results (1961, 60).

Among the advice given to early stereo listeners, one important thing to police the stereo speakers for is to make sure that they are "in phase," meaning that the cones of both speakers "must move forward and backward simultaneously rather than alternately" (Villchur 1961, 60). Several methods of testing to see if the speakers are in phase are suggested, including listening to a monaural record at "a position in front of and exactly midway between the speakers." Villchur continues, "[b]y moving your head slightly you should be able to locate the sound as coming from a definite point between the speakers, if they are in phase. If they are out of phase, the sound will seem disjointed and to surround the listener" (1961). Villchur also offers advice on how to balance the speakers and adjust the tweeters, both of which involve similar adjustments of the equipment and fine tuning of the ear. He ends up on the following note, "[a]fter these controls are set, they should be touched up from time to time as you hear different kinds of records and the newness of the sound wears off" (Villchur 1961, 61). This kind of adjustment of the apparatus and of the ear is characteristic of the discipline required of the audiophile listening subject. Villchur shares with *Understand High Fidelity* a concern with speaker placement. Of such concern, Tony Grajeda says, "[t]he early stereo era was marked discursively by astonishingly specific instructions on where one needed to be stationed in order to achieve the proper effect" (Grajeda 2015, 47). Of the "overriding principle" of speaker placement, he says, "the best position is the position that provides the most natural sound, frees you to the greatest extent from the acoustical environment of your listening room, and brings the sense of openness and space of the concert hall" (Villchur 1961). He makes a joke that makes a point to which this discussion will return, saying, "[o]ne quantitative approach is to take the square root of the area of the triangle formed by two stereo speakers and the midpoint of the listening area, note it carefully, and then have your wife tell you where the speakers sound best" (Villchur 1961, 61). Aside from the gender assumption contained in this bit of humor, it is also interesting in that it sets against each other the needs of the listening subject, the requirements of attaiting "natural" sound, and the absorbing and reflecting surfaces of the décor of the listening space. Grajeda makes the point in his discussion of the "sweet spot" in stereo reproduction that such concern with speaker (and listener) placement is a part of establishing a privileged position for affluent male listeners, saying, "[w]hat stereo is said to accomplish, then is an entire world of 'favored' seats, technologically installing a phenomenology of the 'sweet spot' for every listening subject" (Grajeda 2015, 47). A norm is in the process of articulation in Villchur's introduction to the world of the loudspeaker; a "true" audio "enthusiast" is willing to police himself, his equipment, and his environment for anything that might interfere with his quest for the natural sound of the concert hall.

Complaining wives are not the only potential barriers to the "discriminating ear and [the] fresh memory of the sound of the concert hall" (Villchur 1961, 62). The very nature of the recorded medium can get in the way, and as a discriminating listener you should

> Avoid gimmick records like the plague. You have never heard a harmonica, tambourine, and bongo drum blown up to the volume of a seventy-five piece symphony orchestra because such a thing does not exist in nature, and you therefore have nothing to which you can compare the recorded sound. Also avoid records of electronic instruments and of crooners; their sound has no existence except through loudspeakers, and again there is no live standard of comparison." (Villchur 1961, 60)

Setting aside for the moment the specious nature of the claim that "natural sound" is something that can be achieved through recorded means, Villchur is arguing that human ears need to be trained to recognize the sounds that are natural to human ears, and that it is preferable, or even possible for the medium through which the sound emerges to disappear. He notes that listeners are sometimes swayed by "preconceptions," believing that smaller speakers emit smaller sound and that the "bigness" of the speaker enclosure can equal "bigness" in bass response (Villchur 1961, 62). He advises the listener to turn his back to the speakers in order to combat these preconceptions, because ultimately, "[t]he best compliment one can pay to the speaker is to forget about it and concentrate on the music." (Villchur 1961, 62). What is important for this discussion is to remember that the human ear needs to be trained in order to hear natural sounds, and the sorting out of the natural and unnatural attendant in such training, what Foucault refers to as "the 'physics' of power," disciplinary power that "seems all the less 'corporal' in that it is more subtly 'physical'" (1979, 177).

Foucault notes the importance of establishing "normalizing judgement" to the project of disciplining subjectivity. Coercive pressure to conform to given norms of audiophile practice mostly assumes the form of friendly advice on how to achieve the best results out of one's equipment. Considering audiophilia as a Foucauldian "disciplinary system" requires discussing how normalizing judgment affects audiophile subjectivity. Foucault says that such a system "enjoys a kind of judicial privilege with its own laws, its specific offenses, its particular forms of judgement. The disciplines established an 'infra-penalty;' they partitioned an area that the laws had left empty; they defined and repressed a mass of behavior that the relative indifference of the great systems of punishment had allowed to escape" (1979, 178). Chapter 4 discussed the ways that discursive pratices serve to build communities and individual listeners on the level of language, but this chapter is more concerned with how discourse intersects with bodies in relation to machines and their environments. But, what of the question of how such normalizing judgment is established and disseminated? Advertising and specialty publications such as *High Fidelity* certainly played their part.[5] In addition, dealers and communities of high-fidelity hobbyists also contributed to the disciplinary constitution of the listening subject.

Anderson discusses the part played by the stereo demonstration records made by equipment manufacturers and record companies in educating viewers in the sounds to listen for and the adjustments required to get them. Of such discs, Anderson says,

> Ostensibly, demonstration discs acted as samples of stereo and, as *High Fidelity* put it, were designed "to sell a whole new listening medium" (Smith 1959, 45). One manner of doing this was to explain how the new recording medium

works by displaying a variety of sonic possibilities. The key word is "variety," since, with rare exception, the quality demonstration disc is prized because it verifies a system's capabilities by skillfully exhibiting several musical genres, sonic events, audible locations, and individual styles. (2006, 144)

Sterne makes the point about fidelity that it was, "from its very beginning, about faith and investment in . . . configurations of practices, people, and technologies," and that it required the effort of the listener to discern what was the true and natural sound of music being made in space from the noises both of the environment and the apparatus reproducing the sound (2003, 283). Sterne cites an advertisement for Kolster radios to make the point that, "the ear *needs* aesthetic training," the ultimate project of high-fidelity audiophilia (2003, 281, emphasis his). In order to get the most from the high-fidelity stereo system, the listener had to cooperate and involve himself in the process of making sound, because "[a]lthough the competent recording engineer would 'take into account the probable acoustic surroundings—the average living room in the average home—in which the recording [would] probably be heard,' not surprisingly, a certain amount of responsibility needed to be placed on the listener" (Anderson 2006, 147). This responsibilty included the configuration of the equipment, its placement in the listening space, and the coordination of the listener and the equipment within that space.

At the time Villchur was writing, and continuing until the advent of the CD era and beyond, the audiophile listening subject, it should be added, was to a great degree also the vinyl record listening subject. In an issue of *Hi-Fi Stereo Review* published in 1960, Hans H. Fantel points to what was at stake, saying "the record player is the wellspring for much of the music that flows through the rest of the sound system: and if the music is polluted by distortion at the source, even the very best speakers and amplifiers cannot restore its tonal purity" (1960, 61). Interacting with shellac and vinyl records required and requires special care in at least three points of contact, the listener and the record, the record and the turntable, and the stylus and the groove. A serious miscue at any of these points of contact could very well result in damage to or even destruction of the record or the apparatus used to play it.[6] As many commentators note, Yochim and Biddinger among them, it is the very fragility of the medium that sells it for some listening subjects; "vinyl has always been evocative of people's humanity, fallibility and mortality" (Yochim and Biddinger 2008). Thom Holmes says of the shellac compounds used in making records before the vinyl era, "[e]ven under the best conditions, however, shellac was subject to gross surface noise produced by the abrasive filler—most of it limestone. Another great problem was the brittleness of the record, requiring great care and cost in handling, packaging, and shipping" (Holmes 2013). A 1920 manual for a Victrola XI Talking Machine, while promising, "Victor records are practically indestructible" included the following set of instructions:

> *First*—Keep the record free from gritty dust particles, as dirt on a record has a tendency to make the reproduction scratchy and tends to wear out a record rapidly. A small piece of velvet carpet glued to a wooden block make's an excellent and cheap record cleaner, and should be used on records, before playing.
> *Second*—Keep records in Victor Record Albums, or if no filing device is used, in envelopes in which they are marketed, lying on a flat surface.
> *Third*—Do not pile large records on small ones. Keep same sizes together.
> *Fourth*—Do not expose records to extreme heat.
> *Fifth*—Soap and water or any other cleaning compounds should not be used on the records.
> *Sixth*—When shipping, take care to have records so packed that no sliding of one upon another can take place. This will prevent any scratching which might be caused by hard particles between the records.
> *Seventh*—The use of Victrola Tungs-tone stylus or Victor needles only will prolong the life of records greatly. (Victor Talking Machine Co. 1920)

The instructions are silent on how a "practically indestructible" object might need to have its life prolonged, but they do begin to set up the care and consumption of records as a disciplinary project from the start. As Holmes noted, the fragility of the shellac compounds used in the manufacture of the records in particular render them very difficult to preserve at best. Andre Millard adds,

> [m]uch of the unwanted noise on a 78-rpm shellac disc was the sound of the disc surface being worn away . . . On a typical Victor machine, the downward pressure was estimated to be 50,000 pounds per square inch. A shellac disc lasted only a short time under such punishment; average life was between 75 and 125 plays. (2005, 203)

Although vinyl records are significantly less fragile than the shellac records they displaced, they still require special care in order to obtain optimal results, aside from any particular audiophile considerations. Bartmanski and Woodward note that vinyl "demands attention and ritualizes listening to music. In terms of attention, it is like a guest in one's apartment, meaning that you can't ignore vinyl—playing requires periodic and regular treatment" (2015, 58). And, how you treat your vinyl says a lot about *you*, the listening subject, at least according to the norms that have been set in place regarding the "periodic and regular treatment" of vinyl records.

Particularly in the early development of the vinyl LP as a medium, the special care that it took to play and to maintain was part of its allure. Richard Osborne says that the "care and attention that the LP and its equipment required ultimately helped to ensure its cultural prestige. Not only was the LP too expensive for teenagers, it was also considered to be too fragile, as was the hi-fi apparatus that was produced to accompany it" (Osborne 2012). In fact, Osborne quotes Max Jones writing in *Melody Maker* about the vinyl listener in 1953 as saying, "Good habits creep up on him, and they are fostered by a natural desire to protect the sum of money he has invested in Long Playing records" (Osborne 2012). Some of the early norms established by record labels, dealers, print media, and collectors are still followed more or less faithfully by vinyl listening subjects, such as gripping the record at the edges so as not to put fingerprints on the grooves and wiping the record with a lint-free cloth or soft brush to remove dust before playing them. When such norms are imparted to new or inexperienced vinyl collectors and listeners, it is often in the imperative, such as this bit of advice by Michael Waehner, "I can't say this enough times: you should clean every record before you play it. If you play a dirty record, you can do permanent damage. A clean record is a happy listener" (2012). In the guise of such practical advice are normalizing judgments established, leaving the listening subject to believe that he or she is doing something wrong if not following it.

Foucault says that discipline "has the function of reducing gaps. It must therefore be essentially corrective" (1979, 179). Waehner helps establish such a gap and offers such a corrective in the way records are maintained, saying, "After any cleaning other than routine dusting, always put a record into a brand new sleeve" (2012). No sooner is a norm established, however, than it gets pushed to its farthest extreme. Waehner discusses methods of cleaning from routine dusting and simple cleaning to record cleaning kits and vacuuming machines before an extended set of instructions for "intense cleaning," which Waehner says he adapted from an even more involved article by Michael Wayne originally published in the periodical *The Tracking Angle*, called "Zen and the Art of Record Cleaning Made Difficult." The article Waehner refers to is reprinted on the website Analog Planet, and it provides a dire warning for subjects who do not heed its advice, when Wayne says, "Play a record with foreign matter sticking to its grooves, and you run the risk of welding it in permanently" (Wayne 2012). Wayne offers his advice to the "well informed audiophile record collector," and to call it "involved" would be an understatement. Suffice it to say that the original article was not misnamed, since cleaning a record "properly" involves first obtaining specific types of fluid, cleaning cloths, and brushes, some of which are not easy to get.[7]

The process itself is too involved to summarize effectively, but it involves three complex processes that Wayne refers to as "steps," each of which uses the application of various cleaning fluids and special applicators, such as Orbitrac pads and Nitty-Gritty First solution, to remove not just additional dirt that builds up in the grooves of the record at the molecular level, but also to remove the remnants of the cleaning fluids used in the previous steps. Following these extensive cleaning processes, Wayne recommends not returning the record to its original sleeve, since it is "smelly, oily, and decades old" in most cases (Wayne 2012). Wayne ends with the following bit of advice:

> While the entire process described herein may seem hopelessly complex, in practice it is logical and straightforward, albeit time consuming. Do no take short cuts: they will lessen the effectiveness of the procedure and may lead to the contamination of record cleaning pads and the soiling of all subsequent discs which come into contact with them. (Wayne 2012)

Friendly audiophile advice is often given in a more or less imperative tone of voice in addition to offering dire warnings of the dangers and the potential damage that will result from any deviation on the part of the reader. Foucault says that disciplinary power "refers individual actions to a whole that is at once a field of comparison, a space of differentiation and the principle of a rule to be followed" (1979, 182). Such is the way that the norms of audiophilia and vinylphilia are established and imparted. Resisting these norms, as does the emphatic commenter at the end of the website reprint of Wayne's article, who says, "I'll quit vinyl before going through that process," aside from opening oneself up to ridicule from one's fellow vinylphiles, places the resisting subject in an oppositional relationship to the original imperative, helping to establish it as a norm (Wayne 2012).

Foucault makes the point that even the guard who subjects the prisoners to surveillance cannot escape the normalizing judgment of discipline. The editor of Analog Planet does not exempt himself from such ridicule, posting a microscopic picture of his stylus caked in debris in a blog post, stating, "This is a digital microscope shot of my Lyra Atlas. I am embarrassed" (Fremer 2013). Fremer is embarrassed partially because he operates a website that offers advice to vinyl users, and he has discovered evidence that perhaps he has not been following his own advice as well as he might. At first Fremer sounds defensive: "Now I play either new or cleaned records and I do clean that stylus. I swear! With the naked eye it appeared clean but clearly it is not, even though the stylus tip appears to be" (Fremer 2013). Eventually Fremer comes to acceptance of his irresponsibility in the care of his stylus and considers it to be a lesson for his readers, saying, "if you aren't careful you could end up with this kind of mess. And I thought I was being careful. I'm going to redouble my efforts" (Fremer 2013). At the end of his blog post, in fact, Fremer appears to welcome ridicule for his laxness in the care of his stylus, saying, "it's time for your mocking comments. It's okay. I'm ready, willing and deserving" (Fremer 2013)! Perhaps unsurprisingly, Fremer's blog post does not receive the mocking comments he invites, but instead engenders a lively discussion and debate about proper stylus care and ways of cleaning records so that the stylus does not wind up like the one in Fremer's picture. Having a dirty stylus does, however, become an issue of one's identity as an audiophile, as does the consideration of proper ways to keep it clean.

One must do the best one can; redouble one's efforts if one strays from the path of the pursuit of audio perfection. The best way to maintain a stylus, the proper type of inner sleeve in which to place records, the correct way of cleaning and storing vinyl, the best type of turntable on which to play those records or stylus to track the grooves, whether to use a solid state or a tube amplifier, the most "life-like" speakers and the most optimal types of wires used to connect the components of a sound system, all of these subjects and more are likely to spark impassioned debate among individual listening subjects when they are put together. It might even be reasoned that the community of audiophiles, such as it exists, is built on the foundation of such debates. Should one clean one's brand new records before playing them? Is the Discwasher system the best record cleaner, or are the Nitty Gritty or the VPI vacuuming systems superior at eliciting optimal performance from one's records? What the debates themselves cannot obscure is the question of whether or not one actually needs to do *any* of these things, that they all need to be considered in some way is in fact beyond question. The decisions one makes and the actions one takes are less significant than the fact that one is making decisions and taking actions, and that one's identity as a "true" audiophile hangs in the balance.

That the audiophile identity is to some degree inflected by gender ideology should perhaps not be overlooked. Audiophilia should be understood in this context as a species of technophilia, and as the editors of the book *Gender and Technology: A Reader* note, "the relationship between gender and technology is a reciprocal one. Ideas about society, including gender, shape the ways we make do and design things; these things, in turn, become part of how we identify, structure, represent, and perform gender" (Lerman, Oldenziel and Mohun 2003, 121). The gear aspect of records and the equipment used to play them on was, according to industry historians, one of the things that initially drew men to music, which had previously been marked as "feminine" and unmanly. Ken Rockwell says of early high-fidelity sound, "In those days, you needed to be an electronic engineer just to get sound, much less good sound. We were all about the music, but if you weren't a BSEE (electronic engineer), all you could hope for was a plug-and-play phonograph. It took real men to get high fidelity" (Rockwell 2012). Mark Katz sums up how the phonograph made music safe for bourgeois masculinity when he says that the phonograph

> mitigated the supposed "feminizing" influence of music (particularly classical music), for as a machine it opened opportunities for tinkering and shop talk, traditional men's activities. As one writer explained in 1931, "That men are notoriously fascinated by small mechanical details is a securely established fact. Well, then, is it any wonder that . . . men suddenly became profoundly interested in the phonograph?" (Katz 2004, 66)

As the technology became more complex, its identification with masculinity became ever stronger. Rachel Maines classes high fidelity technologies with ham radio and photography

> as what are called "scientific" hobbies, in that serious practitioners usually need to acquire a fair degree of technical expertise in the physical sciences related to their craft; all three are in the tradition of affluent Western mostly male hobby experimentation that found expression in earlier centuries in amateur involvement with scientific societies. (Maines 2009, 110)

Maines ties the growth of such technological hobbies to the era beginning at the end of World War II, a time in which industrialization had brought about the increases in affluence and leisure time needed to pursue such activities. In turn, entire industries grew to accommodate these hobbies and their technological concerns.

And, make no mistake, audiophilia and vinylphilia, while more or less affordable at the entry level, can become very expensive identities to maintain indeed. For the budding vinylphile, new turntables can be obtained for less than $100 and connected to a laptop computer via USB cables, and a newly committed listening subject can buy an all-in-one unit with amplifier and speakers included for about the same amount. It is assumed, however, in most of the discourses surrounding vinyl and audiophilia that such measures will very soon be found to be unsatisfactory, and that the listening subject will graduate to more involved, and expensive, ways of satisfying the vinyl habit. Coming in at number 6 among the "9 Things Audiophiles Hate" at HDtracks.com is "Entry Level Gear" (9 Things Audiophiles Hate 2014). In fact, the owner of a Crosley all-in-one unit will eventually become the subject of much coercive ridicule designed to get the listener to upgrade to better equipment that will be less potentially damaging to the vinyl records. Of brands for which "serious" audiophiles express an undue antipathy, Crosley is only matched by the name Bose, the manufacturer of expensive—some would say "overpriced"—audio systems that advertise themselves as audiophile systems. The website, Poor Audiophile, outlines the reasons for this antipathy, saying, "audiophiles resent the fact that Bose isn't a peer-reviewed ecosystem. They don't submit their products into trade mags for reviews and independent measurements. To many audiophiles, that means that they have something to hide. Audiophiles also resent Bose's marketing, their proprietary interconnects, and feel that they are simply overpriced" (The Poor Audiophile 2013). The writer concludes, "one thing audiophiles could do is nurture people into the hobby better. There are different ways to guide people into better musical experiences. Bashing other people's gear and their tastes isn't one of them" (The Poor Audiophile 2013). Ken Rockwell makes the point that audiophiles "spend fortunes on the wrong things, which are the high-profit-margin and well advertised items like cables, power conditioners, amplifiers, power cables, connectors, resistors, and just about everything that has almost nothing to do with the quality of reproduced music," a debatable point to be sure, but that there exists such a debate is part of the point (Rockwell 2012).

Acoustic Sounds, an online store catering mostly to audiophiles and vinylphiles, lists as their highest priced turntable the SME Model 30/12 including the Series V-12 tonearm at the princely cost of $35,999.99, and it is by no means the most expensive item of its type available (SME Model 30/12 n.d.). Of course, this price does not include the addition of a cartridge, the most expensive of which, the Koetsu Blue Lace Platinum Magnet, will run the audiophile another $19,950.00, not counting $4,000.00 for the Optional Diamond cantilever (Koetsu—Tiger Eye Platinum n.d.). Multiply those costs by the amount it would take to get the correct amplifier, phono stage, speakers, and power supply that such a turntable would require, and the cost would be approaching $100,000.00 easily. Audiophilia, however, is only partially about ostentatious display and snobbery, but it is also about tactics, techniques, and mastery. Maines notes of hobbies in this regard that they "provide a venue for skill and the pursuit of excellence" (Maines 2009, 123). Developing one's skill in pursuit of excellence places the quest for high fidelity well within the realm of power relations, well within Foucault's body-tool, body-machine complex.

One of the first tasks of the nascent electronics industry, in fact, was getting the making of such tools out of the hands of the amateur and into the open market. Building home audio components was the only way to get them in the post-war years from which hi fi home audio sprang. Steven Stone makes the point, saying, "veterans took advantage of training programs during their time in the service and subsequently through the GI Bill, to learn electronics. Armed with their newly gained knowledge vets began building their own music reproduction systems, and the hobby and business of high end and specialty audio began" (Stone 2011). Once component manufacturers took over, the expertise of the listening subject had to be expressed through the configuration of the home audio system, subjectivity was expressed by the ways in which the equipment and the records were cared for, and thus were the ways in which disciplinary power to express itself. Foucault talks about "surveillance expressed in the architecture by innumerable petty mechanisms" (1979, 173). There one finds "the progressive objectification and the ever more subtle partitioning of individual behavior" (Foucault 1979, 173). The "sweet spot" of audiophile practice provides just such an architecture of imaginary space. Tony Grajeda says, "The disciplinary procedures here for positioning oneself in relation to home audio technology reveal how thoroughly instrumentalized the listening experience had become by the mid-century era of high fidelity. What also stands out about this sheer rationalization of an aural field centered on a lone auditor is the degree to which stereo consistently operated across an economy of vision and visuality" (Grajeda 2015). Audiophilia, perhaps paradoxically, becomes panoptic in its power in this sense. Ken Rockwell notes,

> Since sound and music perception is entirely imaginary (you can't touch or photograph a musical image), what and how we hear is formed only in our brains and is not measurable. Our hearing therefore is highly susceptible to the powers of suggestion. If an audiophile pays $5,000 for a new power cord, he will hear a very real difference, even though the sound is unchanged. (Rockwell 2012)

The money one spends, the actions one takes, the habits one develops in pursuit of this idealized space contribute to the building of a world, much in the same way that Desmond on *Lost* built his world around the input of the numbers, and whatever its truth or reality, it becomes very real to him. That the "sweet spot" ultimately resides inside one's head makes it no less a spot for the listener, and training the ears to chase after it no less demanding, and this training and these demands contribute to the constitution of the listening subject.

NOTES

1. Lindelof, David. (Writer), Bender, Jack. (Director). September 24, 2005. "Man of science, man of faith" [*Lost*]. In J.J. Abrams, Damon Lindelof, Carlton Cuse (Producer). Los Angeles, CA: Bad Robot, Touchstone Television, ABC Studios.

2. According to Lostpedia, there are various theories to explain what disease the vaccine is meant to combat, but by the end of the series there are no definitive answers (Vaccine 2014).

3. Godley and Creme, "I Pity Inanimate Objects" in *Freeze Frame*, Polydor Records, Man-Ken Music, Ltd. 1979, LP.

4. Jonathan Sterne cites this same set of instructions in a different context, discussing how it taught new users how to recognize differences in tone and volume (2003, 268–269).

5. Rachel Maines notes in this regard, "Like the literature of ham radio, that of audiophilia was born in the 1920s, with a meager two titles, one of which was the serial Gramophone, which began publication in 1923. Like that of their radio counterparts, the expansion of published titles in the subject areas "Stereophonic Sound Equipment" and "High-Fidelity Sound Systems" occurred in the 1950s and 1960s, with more than 150 titles, including more than 30 new serials, in each of these decades. This literature, however, apparently reached its peak in the decade 1971–80, with 344 titles, of which 48 were new serials" (Maines 2009, 111–112).

6. The author learned this lesson not for the first or probably the last time in 2014 when he dropped a valuable record onto the turntable destroying it and a 250 dollar stylus in the act.

7. For instance, Wayne recommends using triple distilled water, since improperly distilled water might "leave residue of noise, depositing minerals in the grooves, defeating the entire purpose of this entire, time-consuming process" (Wayne 2012). The problem? The editor of Analog Planet, Michael Fremer, offers the following postscript, "Triple distilled water, available at some pharmacies, is not easy to come by: when I tried to obtain some in NYC, the pharmacist accused me of being either an abortionist or an intravenous drug user. In fact, it requires a doctor's prescription in New York" (Wayne 2012).

Chapter Seven

"The Vinyl Anachronist"

The Role of Social Media in the Formation of Communities of Vinyl

If in fact the "dynamic landscapes" of virtual reality and cyberspace "offer immersion in simulations that blur fact and fiction" (Opie 2008, 5), how are we to locate a space in which the blog *Records My Cat Destroyed* exists and is able to articulate itself? The header image of this blog features a photograph of a shelf of LP records that appear to have been clawed on their spines superimposed over a photograph of an empty cat bed on which rests the figure of a small plush smiling pig cat toy. The blog explains itself as follows:

> I once had a record collection.
> I once had a cat.
> They were not compatible. (Records My Cat Destroyed 2015)

The blog began in September of 2014, and it is still being regularly updated as of this writing. It consists of photo entries of individual record covers perched on the cat bed easel, inner-, outer-, and gatefold sleeves, each of which, with one exception, features the pig posed as if reacting to the record cover, along with a brief description of the album in question. The one exception to the pig toy interacting with the records is the Pink Floyd album *Animals*, which, probably because it features the songs "Pigs" and "Pigs on a Wing," substitutes a plush flying insect cat toy interacting with the album cover. The blogger includes an "Ask Me Anything" link, and a question about both the blog in general and the pig toy in particular elicited the flowing response from a man named Felix: "The stuffed pig ('Roger') was bought years (decades?) ago at Ikea, because it reminded me of the Animals album by Pink Floyd. When I started Records My Cat Destroyed it was rather impulsive, but I wanted some links to the cats in the title."[1] He notes that the pig is named "Roger," perhaps after Pink Floyd's singer, bass player, and songwriter Roger Waters. The blogger appears to have a very large and varied record collection, mostly centering on rock and popular albums of the 1970s, and his cats appear to have done quite a bit of damage to his record collection, if only the sleeves. The blog appears factual, and it blurs very few lines other than the lines of whether or not a given topic is interesting to a blog reader, but the facts that it exists at all, that it has a dedicated readership, and that it features vinyl records are all worthy of remark. The blogger shares quite a bit about his tastes and interests in this blog, but very little about himself, except by implication. It is, nonetheless, one of many varied ways in which social media disseminates information and entertainment regarding vinyl records and the technologies on which to play them. This chapter will argue that part of the reason vinyl has made its comeback and continues to grow in popularity is the communities of interest it creates and the interactions among vinyl aficionados it engenders, and that the Internet and social media have played a significant part in the proliferation not only of selves, but of selves in relation to vinyl records.

Perhaps the most remarkable about being a vinyl listener and collector in 2015 is the fact that being a vinyl listener and collector *is* remarkable. Not that people didn't listen to or collect vinyl records in the past, even obsessively so and therefore worthy of remark. No, the reason listeners and collectors in 2015 are worth remarking might have more to do with the fact that vinyl records were declared an obsolete technology by the record industry and mainstream culture at least twenty-five years ago. John Silke makes the case very well and raises a significant question when he says,

> Vinyl. Should be obsolete by now shouldn't it? Consigned to the museum of Twentieth Century artefacts alongside typewriters, telegrams, 35mm film Walkmans and Pac Man. In 2012 we can carry in our pockets iPods, iPhones and the like, that can hold the equivalent of a dozen record collections. So why on earth would anyone still spend time and money on something so antiquated that takes up so much space? (Silke 2013)

And, why on earth people would either continue to use vinyl or even return to vinyl under such circumstances is not the only question worth asking. Earlier chapters of this book have already dealt with the question of why listeners have either stuck with or returned to vinyl. Who kept the vinyl flame burning during the dark years of the 1990s and 2000s, and why? This chapter also deals with the question of how so many people returned to vinyl at roughly the same time, and in particular examines the role that the Internet has played in this return, the communities it has helped to build, and the ways that it has helped to spread the word.

Yochim and Biddinger locate the end of the (first?) vinyl era in 1983 with the introduction of the compact disc. There wasn't much of a battle, since as of 1983 the cassette tape had already surpassed vinyl in overall sales, "cassette sales totaled nearly 237 million, surpassing the 209 million units that vinyl accounted for" (Christman 2007). Yochim and Biddinger note, "The last proclamation of vinyl's death in *Rolling Stone* appeared in 1988, and only five years later, in 1993, articles announcing vinyl's return emerged" (2008, 187). Aside from audiophiles, those with systems worth 50 or 100 thousand dollars who believed that CDs did not sound as good as vinyl, and aside from jukeboxes that still accounted for the lion's share of the market until at least the "mid-aughts," vinyl was mainly kept alive by hip-hop crate diggers.

Even if vinyl never truly died, for a time it became harder to acquire. Todd Souvignier makes this point when he says, "[by] the 1990s one could no longer buy a turntable or LP at any major retailer and most people seemed fairly content with that. But one group held onto their records and phonographs: DJs" (2003, 42). In a discussion of what kept vinyl alive, *Spin* magazine talks to Martin Imbach, of Georgetown Records in Seattle, Washington, and says,

> "I think the hip-hop, drum-and-bass, jungle, etc. scenes kept vinyl going strong during those years," he writes, speaking from experience as a collector, as the store didn't open until 2004. "It seemed like alt-rock bands and labels jumped ship for the most part. The exception was the burgeoning underground garage and Punk scenes—those bands often eschewed CDs completely, and still do." (Hogan 2014)

Souvignier discusses the attraction of vinyl for club and hip-hop DJs in the years after the introduction of the CD, saying, "DJs need to cue songs, beginning them at specific points. The first several generations of CD players were so simplistic and cumbersome that cueing was easier with vinyl. Some DJs liked to tempo match, adjusting the speed of one record to match another; that technique was impossible with CDs until fairly recently" (2003, 42). Furthermore, for hip-hop DJs, "techniques like scratching and cutting just couldn't be done with CDs" (2003, 42). While such practicalities might account for some subcultures holding on to a format that was largely declared obsolete by mainstream culture and the recording industry, they do little to explain how the format arose from its grave.

In her discussion of the first exhibitions of the Edison tinfoil phonograph, of the fact that souvenir pieces of the used tinfoil were given to and prized by the public attendees, Gitelman notes that "[l]ike all souvenirs, they were belongings that vouched for belonging. They were artifacts that vouched for facts. Publicly made and privately held, their very material existence offered a demonstrable continuity of private and public memories, hard evidence of shared experiences" (2008, 39). Surely she could also be referring to vinyl records in this description. This chapter argues that while the vinyl flame was kept burning by audiophiles, club, and hip-hop DJs during the dark years of digital music, it is no accident that the reemergence of the vinyl format roughly coincides with the rise of the Internet and online modes of socialization and communication. Experiences are still shared among vinyl listeners, but to a significant degree, they are now shared online through social media. Before taking an ethnographic look at one particular vinyl-related online community, this chapter will discuss the ways in which social media have influenced the adoption or the re-adoption of the vinyl and other analog formats.

It is probably close to a cliché at this point to say that the Internet has had a profound impact on the ways people socialize and the cultures they create, but it is still worth saying in the context of the reemergence and continued vitality of the community of vinyl users. Digital technologies are changing the ways music is conceived of and listened to in significant ways. As Mark Katz remarks, changes in the ways people exchange, think about and interact around music,

> could only take place because of a transformative web of technologies that, at their root, enable all information—including music—to be represented, stored, and distributed as long strings of 1s and 0s. This is digital technology, and it is bringing about what musicologist Timothy D. Taylor describes as "the most fundamental change in the history of Western music since the invention of music notation in the ninth century." (Katz 2004, 159)

While Katz's main subject is the ease with which Internet users can exchange and discuss digital music files, a case could similarly be made for the ways in which the Internet is helping users acquire and talk about vinyl records. Mark Weinstein of Amoeba Records has been quoted as saying; "I see amazing potential in the digital realm. Create a museum online and make its pieces available for purchase" (Calamar and Gallo 2009, 181). Minus the profit motive, Weinstein is describing just what many vinyl evangelists do on websites, blogs, and other social media. That said, the auction site eBay preceded Discogs by many years as an online source for used vinyl, and it thereby helped to keep vinyl alive for many users who had little other recourse if they wanted to buy vinyl records. Thus, the profit motive helped to keep vinyl collecting going up to the point where it could be picked up on by other social media, such as tumblr, Facebook, Twitter, and blogs. The profit motive also plays a part in the proliferation of online subscription vinyl services of the type discussed in chapter 3, with Vinyl Me, Please, Vinyl Moon, and Turntable Kitchen, among others, joining the aforementioned Feedbands,

Third Man Records, and Light in the Attic to offer limited edition records on a monthly basis (Hickey 2015). In addition, as 2015 drew to a close, Columbia House, a monthly subscription service that bills itself as a "record club" and goes back as far as 1955 (Columbia House n.d.), announced that it is reintroducing vinyl records as of 2016 (Fingas 2015).

One online vinyl evangelist, a blogger who gives this chapter its title, goes by the name "The Vinyl Anachronist," a name under which Marc Phillips has been writing about both digital and analog audio and vinyl records since 1998, according to his blog profile. He often redirects blog readers to a regular column he writes for the online music magazine *Perfect Sound Forever*, at which site readers can find his first column under his online name. Back in 1998, Phillips took note that

> The only musical format that has increased its unit sales over the last three years has been vinyl LPs. Let me repeat this: THE ONLY MUSICAL FORMAT THAT HAS INCREASED ITS UNIT SALES OVER THE LAST THREE YEARS HAS BEEN VINYL LPS [*sic*].
> You think that I'm crazy. I've certainly been called that over the last few years every time I even faintly suggest that the vinyl LP, an analog medium, sounds better than every one of its digital counterparts, including DVD. But the statistics remain. These are not my statistics, but those of Billboard, and many others. And the reason for this "Vinyl Renaissance," as Stereophile writer Michael Fremer calls it, is because LPs sound better than CDs. (Phillips 1998)

After calling attention to the irony of making this claim in an online publication that takes its name from Sony/Phillips's promotional motto for the introduction of the compact disc, by telling the story of his own transition to digital and then back to analog, Phillips makes the case that if the listener has a good turntable and well cared for records vinyl sounds better and "less fatiguing" than CDs. Phillips also notes that he is not alone in this change, saying that many people have made up their minds that vinyl is superior, "as witnessed by the aforementioned 'Vinyl Renaissance.' People realize that music has been missing from their lives and maybe digital is the cause. More people are bringing their old ARs and Garrards and Duals and Aristons out of their attics than ever before. And they're diggin' it" (Phillips 1998).

It can be said that a "discourse of sound fidelity" sanctions and authorizes Phillips's assertions, about which Jonathan Sterne says, it "is as much a product of and a player in cultural history as are the machines that it purports to describe. The possibility that a reproduced sound could be faithful required that listeners and performers have faith in a network: a set of social relations, technologies, and techniques" (2003, 283). As a medium, vinyl records and the analog gear on which they are played are subject to Lisa Gitelman's definition of "socially realized structures of communication, where structures include both technological forms and their associated practices, and where communication is a cultural practice, a ritualized collocation of different people on the same mental map, sharing or engaged with popular ontologies of representation" (2008, 7). However, what if the "mental map" in this case traces the topography of cyberspace? To be sure, discursive practices around the collecting and consumption of analog records did not originate with the information age. From the very start of the history of recorded music, Roy Shuker points to the importance of record appreciation societies, saying that they were

> important for fostering collecting as a selective activity, involving the acquisition of discrimination, discernment and cultural capital. initially, this was especially the case in relation to classical music and, in particular, opera singers. later, jazz, while originally a populist genre (the Jazz age of the 1920s), proved increasingly attractive to those looking to explore a more musically complex style of popular music. (2010, 22)

In addition, there is the entire history of the music press in providing a helping hand in the dissemination of vinyl discourse, which Shuker calls "a key part of the infrastructure of record collecting" (2010, 137). What has the intervention of the Internet done that already established modes and methods of approach could not already do?

Aside from the fact that it has increased and eased the access to information of those who can afford or otherwise procure the technology it requires, in his discussion of its online libraries, Martin Hand says that digital culture has "become the vehicle through which to reactivate earlier ideas associated with library postmodernization (interactivity, empowerment, and participation), framing the web as somewhat deterministically empowering information and decentralized learning technologies" (2008, 97). Hand makes two related points that resonate with the current discussion. Calling it "the dominant narrative of digital culture," first, he says that the term "access" can be defined as "a democratization or 'flattening' of culture, of new cultural spaces and forms which are inherently more accessible than ever before because of the place-defying structure of digital communication technology" (Hand 2008, 75). However, Hand says that access can also refer to "new disconnections alongside connections, to new territories and zones, and to divides between them which need to be 'bridged' as select physical environments become informatized to an unprecedented degree" (Hand 2008, 75). Hand concludes that regarding social and public policy around the Internet, "the democratizing qualities of digital culture must be delivered to citizens by overcoming social barriers and digital divides" (2008, 75). And, the term "access" assumes a new tonality in the age of social media, which introduces a whole new level of constant communication via laptop, tablet, and smart phone. In a discussion of the "always-on lifestyle," the mode of interaction engendered by social media is defined as being "not just about consumption and production of content but also about creating an ecosystem in which people can stay peripherally connected to one another through a variety of microdata. It's about creating networks and layering information on top" (boyd 2012, 75). Socializing, interacting, creating and responding to content may not have been created by the Internet, but they certainly have been inflected by being channeled through cyberspace, and, furthermore, all of this has had more than a little to do with the reemergence of vinyl record listening and collecting in the last twenty or so years.

In the first place, it is probably worthy of note that the collection of material things, even to some degree anachronistic things, is being largely supplemented and facilitated by the virtual culture of cyberspace. David Banash refers to "the great paradox in our contemporary moment, the fact that

> although there is ever more depth, quality, and scope available for free in the virtual world of the Internet, many everyday objects available to ordinary people at reasonable prices have become noticeably less well-made. Somehow it almost seems as if our newly found virtual abundance were being paid for by a marked impoverishment of everyday physical objects, particularly the most ordinary kinds of objects. (Banash 2013)

Banash's conclusion could be talking more or less about vinyl reissues when he says, "[g]iven the rise of virtual environments and the cheapening of material everyday life objects, some consumers are turning to reproductions of a lost ordinary that supports a profoundly emotional and embodied experience, though that experience now is really the commodity being sold, and the object is, in some sense, a kind of self-conscious prop that enables it" (Banash 2013). However, Banash is only telling part of the story of collecting and its relationship to cyberculture.

D. Robert DeChaine approaches the part missing from Banash's assessment more closely when he says, "I assert that our collecting habits also include an attention to matters of culture, social values, and self-presentation" (DeChaine 2013). In his discussion of social media and the culture of reading, more specifically the "social networking site Goodreads," Matthew James Vechinski makes the point that a book "is an example of a representation that blurs object and experience," a claim that can also certainly be made for vinyl records (2013). In fact, Veshinski could also be describing what the website Discogs does for listeners when he calls Goodreads an "aggregation of information about reading activities [that] creates for users a virtual book collection" (2013). On the Goodreads site, "[u]sers assemble an online library of book titles they are reading, have read, or want to read, and classify their selections with customized tags" (Vechinski 2013). Discogs (http://www.discogs.com) offers its users a similar service for their record collections, except that to the curatorial and social aspects of displaying one's collection is added the ability to buy and sell records.

At this point it might be worth noting what David Hesmondhalgh says about the social aspects of music, highlighting "music's enhancement of *our sense of sociality and community*, because of its great potential for providing shared experiences that are corporeal, emotional, and full of potential meanings for the participants" (2013, 48). Both Goodreads and Discogs sites include both personal profiles that allow users to share information about themselves and their collections and discussion forums related to either books and reading in the former case or records and music in the latter. Vichinski draws a distinction between collecting and curating, saying that while "the term accurately describes the addition of information to social networking profiles, curating privileges the transaction, the moment of conscious choice, while theories of collecting more fully address the dynamics of arranging sets of items over time" (Vechinski 2013). Social media allows users to make those conscious choices a matter of public record, offering opportunities to build communities with like-minded individuals who share the user's tastes. Vechinski points to the user profile as a means of self-presentation and of socialization, saying,

> emphasizing the personal profile as a record of all the activities performed by a user, Goodreads highlights curating as well as collecting. Each action is a choice, and the sum of the choices represents the user and his or her collection of books. Other users regard the profile as a document of taste, and they form relationships with users whose judgment they admire. This focus on the profile establishes a resemblance to the self-fashioning and networking capabilities of sites such as Facebook. (Vechinski 2013)

Hesmondhalgh, however, sounds a note of caution about the "competitive individualism" of these types of display as they relate specifically to music, declaring that it "can be part of status battles to show one's openness to a variety of lifestyle pleasures and one's superior emotional range" (2013, 43). To some degree, "music might be the basis of status battles in modern society," as individuals display the superior tastes, emotional capacities, and sensual pleasures of their collections in online forums such as Goodreads and Discogs or on social media such as Facebook and Instagram (Hesmondhalgh 2013, 43).

Vechinski separates sites such as Goodreads from the more mercenary profit driven e-commerce sites such as Amazon.com, because Goodreads does not solicit buyer reviews and it tracks the process of readers through the books they are reading in the present moment. However, the self-fashioning and the networking aspects of user reviews and comment sections should not be underestimated, since along with user profiles, discussion forums, comment sections, tweets, and blogs they constitute the power of social media.

Jay Rosen makes a closer approach when he talks about "The People Formerly Known as the Audience," saying to purveyors of traditional media, "[t]hink of passengers on your ship who got a boat of their own. The writing readers. The viewers who picked up a camera. The formerly atomized listeners who with modest effort can connect with each other and gain the means to speak—to the world, as it were" (2012, 13). Rosen quotes Mark Thompson of the BBC as calling social media users, "The Active Audience ('who doesn't want to just sit there but to take part, debate, create, communicate, share')" (2012, 14). It should be pointed out that this active audience is made up of individuals, all of who are in the act of carving out a public place for themselves in this cybersea of information.

It is also worth noting in the context of the current discussion this is more than merely an act of individual self assertion, that "[n]etworked technologies allow us to extend our reach, to connect across space and time, to find people with shared interests and gather en masse for social and political purposes" (boyd 2012, 75). boyd makes the point that behind such acts of self-assertion, "[w]hat I want is to bring people and information into context. It's about enhancing the experience." (2012, 74). Heidi J. Figeuroa-Serriera discusses how Sherry Turkle's research "has fluently described the ways in which in the virtual setting the subject reproduces not only what she or he is but also what she or he would like to be and what she or he would not want to be in the real physical world" (2006, 101). Self-fashioning and identity formation online in the world of social media, then, is a process of

> [a]ssuming a particular identity requires taking on at the same time a universe of meaning. This, then, implies an ethics and aesthetics of the social relationships that have shaped that identity, through multiple identification processes. Self-referentiality then expands inside and outside the screen, inside and outside what we generally call "the person." (Figueroa-Serriera 2006, 102–103)

Of course, the "extended reach" of social media does not go without its naysayers. Of this "new regime" of always-on communication, with its crowded public spaces full of people attending only to their mobile devices, Turkle says, "being alone can start to seem like a precondition of being together because it is easier to communicate if you can focus on your, without interruption on your screen" (2012, 155). In this new regime, "a train station (like an airport, a café, or a park) is no longer a communal space but a place of social collection: people come together but do not speak to each other. Each is tethered to a mobile device and the people and places to which that device serves as a portal" (2012, 155). Whether these new means of socialization make up for the loss of sociability in the here-and-now, it must be acknowledged that they do provide at least some of the stuff of who we are. Turkle makes this point when she states that over the course of our lifetimes, "we never 'graduate' from working on identity; we simply rework it with the materials at hand. From the start, online social worlds offered new materials" (2012, 158). Social networks highlight the social aspects of identity formation. We are not coming into being as ourselves in the world of social media so much as we are coming into being as the selves that we wish to represent ourselves to the world. As Turkle says, "on social networking sites such as Facebook, we think we will be presenting ourselves, but our profile ends up as somebody else—often the fantasy of who we want to be" (2012, 153). And, as go our identities, so also go the communities that they come together to form, at least according to some experts.

When this chapter borrows the term "self-fashioning" from Vechinski in relation to social media, it means the reader to understand the term more or less as Stephen Greenblatt used it, as "an increased self-consciousness about the fashioning of human identity as a manipulable, artful process" (2012, 2). Greenblatt points in his study to "the interpretive constructions the members of a society apply to their experiences" (2012, 4). "Social actions," Greenblatt says in his defense of not treating history and literature as separate discourses, "are themselves always embedded in systems of public signification, always grasped, even by their makers, in acts of interpretation" (2012, 5). Nancy K. Baym makes the point that what is exciting about new forms of social media "is that they allow us to communicate personally within what used to be prohibitively large groups," which "blurs the boundary between mass and interpersonal communication in ways that disrupts both" (2013, 13–14). Baym further notes that "the boundaries between production and reception of mass media are blurred as well," a topic to which this chapter will return. Baym makes a point about social networking communications technologies that could just as easily apply to vinyl records technologies; she says, "the messages we communicate about technologies are reflective, revealing as much about the communicators as they do about technology" (Baym 2013, 29). Bartmanski and Woodward say of things in general and vinyl in particular, "the thing of human design becomes socially significant not only through semiotic and discursive contextualization but crucially through pragmatic contextualization, i.e. specific actions for which it was cut and particular experiences that it affords and effectively fosters" (2015, 67). The question that all of these theoretical notions could raise at this point is, when one presents oneself online through social media specifically as a vinyl person, what does one communicate about oneself in addition to what one communicates about the analog technology through the digital technology?

With these ideas in mind, it might be worth examining a particular form of self-fashioning in response to vinyl records, in the form of a blog begun in February of 2014 and continuing at least up until April of 2015 called, *My Husband's Stupid Record Collection*. The blogger, Sarah O'Holla, states her intentions in her first post; she will be listening to her husband Alex Goldman's collection of "about 1500 records" in alphabetical order by artist's name and posting blog entries in which she reacts to each one (O'Holla, What is this? 2014). O'Holla says that when she first broached the idea her husband said "reading a reaction from a person like me, rather than a person who knows about the history of what I might be listening to, who has been listening to the same stuff for decades and has the vocabulary to talk about it, will be funny, sincere and maybe even thought provoking" (O'Holla, What is this? 2014). She also includes a picture of herself posing with each record as she discusses it. For readers of the blog, the subjectivity of the author is inextricably tied to the objects she holds in her hands. Bartmanski and Woodward note that "[i]n a digitally altered world, the seemingly old-fashioned thingness and unique pragmatic qualities of the vinyl record could and did become somewhat conspicuous and unique" (2015, 68). The self presented by O'Holla in relation to her husband's records serves a two-fold function. First, her seemingly naïve, bemused, and semi-incredulous approach to the records she describes, according to Amanda Hess, "subtly mocks the arcane knowledge that is deeply important to a certain segment of people (music nerds) but is nevertheless inscrutable and potentially annoying to those who are forced to coexist with it (chiefly, their spouses)" (Hess 2014). Second, her unfamiliarity with her husband's vinyl records models a specific type of perhaps gendered response—Hess calls O'Holla's approach to the records "a blend of condescension and wonder" (2014). At the end of her first blog post, O'Holla sets the following rules for herself: "Start with the 'A's' these records are set up in alphabetical order by artist. Listen to the entire thing even if I really hate it. And make sure to comment on the cover art. Are you with me? Let's see how far I can go" (O'Halla 2014). She will be telling a story in addition to reviewing the records. Thus the entire blog sets itself up as a narrative, each individual post making up its own chapter, although the question might arise whether this will be a story of growth and what exactly the blogger will learn in her encounters with the records.

Allison Cerra and Christina James say, "[t]echnology will never define us, but it certainly allows us to express who we are. For some, the physical world may not always accommodate the 'trying on' of different personas, if not the exploration of deeper aspects of one's personality" (2011, 169). O'Holla of *My Husband's Stupid Record Collection* appears to be doing both. She describes her tastes in her first entry as "eclectic on the snobbier side," similar to her husband's tastes, however, because her own tastes are more limited, "in a much more clueless way." Hess notes the women music writers have "pushed back against the blog's conceit," saying that while "[p]laying up her own naiveté allows O'Holla to heighten the contrast between her disinterest and Goldman's obsession in pursuit of humor, [. . .] it also places her in the tenuous position of the ignorant wife" (2014). What does it mean, however, to enact this narrative on a virtual stage? Hess says that the blog can be seen as "part of a time-honored tradition of poking fun at your spouse's eccentricities, and that's a premise that will persist regardless of the cultural state of gender relations" (Hess 2014). What if the reader was to take O'Holla's journey at face value, rather than as enacting stereotypes of female cluelessness in the face of male "expertise?" What if in addition to making fun of her husband's obsession, O'Holla was also attempting to learn something about records, and to teach her readers as she goes along? Highlighting the fact that the term "virtual" can mean "potency and potential" in addition to "almost but not quite," Debra Ferreday notes that "more optimistic theories of virtual reality are hence constantly engaged in playing with the boundaries between the real and the fantastic, between what is possible and the sense of incompleteness, of 'not-quite'" (2009, 40). In the case of O'Holla's blog, the virtual is very much invested in the engagement of the blogger with the real and actual physical objects in the form of the records to which she is listening. The question is whether the "authenticity" presented in the case of these encounters only includes tired gender tropes surrounding female and male marital relationships? Hess notes that not long after the blog debuted, people (mainly male) began sharing it in such a way that made it seem as if the joke of the blog was on the clueless wife. Kate Koepple notes that her reason for seeking out O'Holla in the first place was that she is a woman attempting to enter the "SERIOUS boys club that is the music-industry-audiophile-vinyl-collecting world," saying, "Sense of humor intact, Sarah isn't going to let the boys club keep her from having feelings or writing about them—she reviews music as she experiences it, which is you know, sort of the whole point of music . . . experience" (Koepple 2014). Either way, at least for this blogger, the rewards of such self-presentation clearly outweigh the risks of being subject to readers' interpretations of her motives.

Cyberspace and virtuality both ask and answer these questions about human identity in relation to the objects around us. For Slavoj Žižek, in the cyber age, the question of the self in relation to the "big Other" has become "Am I a machine (does my brain really function as a computer) or a living human being (with a spark of spirit or something else irreducible to the computer-circuit)" (Žižek n.d.)? Žižek discusses Sherry Turkle's response to the idea that cyberspace and virtuality raise questions about human identity in relation to the machine, saying,

> our reaction to this question goes through three phases: first, the emphatic assertion of an irreducible difference: man is not a machine, there is something unique about it...; then, fear and panic when we become aware of all the potentials of a machine: it can think, reason, answer our questions...; finally, disavowal, i.e. recognition through denial: the guarantee that there is some feature of man inaccessible to the computer (sublime enthusiasm, anxiety...) allows us to treat the computer as a "living and thinking partner," since "we know this is only a game, a computer is not really like that." (Žižek n.d.)

Žižek discusses Searle's Chinese Room thought experiment and its conclusion that "a computer cannot really think and understand language" to make the significant point that, "since there is the ontological-philosophical guarantee that the machine does not pose a threat to human uniqueness, I can calmly accept the machine and play with it" (Žižek n.d.). Since there is little threat to it at stake online, according to Žižek, cyber subjectivity is freed up to play with alternative subjectivities, because "you can relax, you are delivered of the burden to be what you are, to live with yourself and to be fully responsible for it" (Žižek n.d.). On the other hand, the online persona I create for myself might be "more myself than my 'real-life' person (my 'official' self-image), insofar as it renders visible aspects of myself I would never dare to admit in RL" (Žižek n.d.). However, the question of the self in relation to authenticity is answered in the virtual world, though, the fact that it is a question is what is important.

Chapter 3 dealt with the idea of "authenticity" as a productive philosophical trope, the discourse on which it is based allowing for new and different notions of self and community in the face of the potential diffusion of digital culture. John Opie calls cyberspace "a seductive alternative existence" that he defines as "Third Nature" as opposed to the "First Nature" of the natural world and the "Second Nature" of the actions of human beings to alter nature. In displacing space and time "as nothing has before it," cyberspace, according to Opie, "provides profound alternatives to First and Second Nature because it defines itself spatially with direct equivalencies of space, place, and time" (2008, 7). Opie discusses ways in which cyberspace is altering the physical space of the living room saying, "'Family' is redefined. Philosophy professor Albert Borgmann identifies a 'self-imposed ghetto of tastes.' The future promises more boutique entertainment. Cyberspace 'can put you in touch with lots of people, but they're all your kind of people.'" (2008, 8). Of course, in Turkle's terms, we could say that the people we connect with online have more in common with the person we imagine ourselves to be than the person we might actually be. Opie makes a point worth considering in the current discussion when he says that human beings have always been creating virtual realities in their minds, defining "cognitive mapping" as "an 'inner space' that contains an internal geography. Our internal landscape is not helter-skelter, but our investment in an ordered mental image" (2008, 26). Landscape is something that we impose on the world as we process the spaces around us through our senses. In other words, virtual reality is not only not entirely new, but it is something that we are always already creating in the interactions of our consciousness with the world around us. Opie also sounds a warning, in his view of "Virtual America," the flaw "is that it remains passive, mechanical, and impersonal, offering a bland, surface existence. It lacks any intrinsic imagination" (2008, 34). However, calling for skepticism of virtual reality on the basis of its lack of correspondence to the "real world" is problematic even on Opie's own terms. What the film that Opie discusses, *The Matrix*, represents is less a warning about the possibility of a "terrible confusion between reality and virtual reality" than it is an allegory for what we are already doing every day of our lives (2008, 34). To conclude as Opie does that "[f]or American culture to prosper, it must have authentic roots" begs the question of how one defines "authenticity" to begin with. Cyber reality hasn't created a new problem so much as it has brought an old problem into stark relief. The binary opposition between the "escapism" of virtual reality and the pragmatic acceptance of the "real world" is specious at best. Ferredey counters this argument using the Freudian notion of fantasy in which "the term materialises as a means of questioning the assumption that dreams and stories are opposed to material 'reality'" (2009, 38). Her argument highlights the role of fantasy in the creation of online communities and subjectivities, saying, "the notion of fantasy as escapism is useful if we consider online reading, not as escape-from (the 'real world') but as escape-into or engagement-with (the text)" (Ferreday 2009, 38).

Ferredey calls the role of fantasy in the creation of our life world "psychical reality," and she quotes Žižek as saying it "is not simply the inner psychic life of dreams, wishes, etc., as opposed to the perceived external reality, but the hard core of the primordial 'passionate attachments,' which are real in the precise sense of resisting the movement of symbolisation and/or dialectical mediation" (2009, 49). Ferredey posits a sense of online community in which fantasy and reality interact in new ways offering "the ability to make the psychic visible and intelligible through textual practices, that by writing the self online it is possible to create communities that are founded on the 'spectral hard core' described by Žižek" (2009, 49). As a mere joke, *My Husband's Stupid Record Collection* would be difficult to sustain, perhaps raising the question of whether or not there is a point to O'Holla's blog beyond making fun of her husband's vinyl collecting habits? In an interview about the blog, O'Holla answers the question, "Why vinyl?" in the flowing way, "I think my love for stories comes before my love for music. Physical records that you can hold in your hands and hear pop and crack on the stereo, that were most likely owned by someone else before they were owned by you, have a story behind them, and I love that" (Koepple 2014). The records provide a connection to the past not only of O'Holla and Goldman, but also to the previous owners of the records. By her eighth post, on *Prince Charming* by Adam and the Ants, Goldman notes that having previously considered the album "a minor entry in the Adam & the Ants ouvre [*sic*]," listening to the record with her has "been a real reassessment" (O'Holla, Adam and the Ants "Prince Charming" 2014). Dialog, pictures, and description combine to paint a picture for the blog reader of an authentic learning experience occurring alongside the humor. Assessing and reassessing the value of these objects is not only an exercise in gender and obsessive interests, it is also an attempt to share feelings with and about actual objects, even though at times the thingness of the records is sometimes not at the forefront of the blogger's interests. When the thingness interests her, however, usually it relates to the stories she talked about. For instance, O'Holla speculates about the check marks a previous owner had placed next to each of the song titles on the back cover of the Chi Lites album,

> What does the check mark mean? It's next to every single song. Was it some kind of superior rating system and only the great songs got a check, and before he knew it, he'd given every song on the album a check? Or did he just like to check it off after he'd heard it, like a to do list. If that's the case, it was like he was just trying to get through the thing—I could have totally used a check mark system while listening to Trout Mask Replica. So what does it mean, this is a check plus album or a to-do list album? (O'Holla, The Chi Lites S/T [1973] 2014)

Answering the question seems a lot less important to O'Holla than asking it. Asking it and speculating about it takes the blog in a different direction from a wife making fun of her husband's record collection, however.

Appropriately, even if coincidentally, O'Holla's very first review is a record called *Song Hits of 1939*, a record she celebrates for its nostalgia; in fact, she discusses it in relation to World War II and her grandmother, whose picture from 1939 she includes, because she feels "it embodies the music on the record a little more" (O'Holla, Song Hits of 1939 2014). This entry is also one of the few, aside from sleeve art, in which O'Holla makes note of the thingness of the thing, the actual vinyl record she is describing. O'Holla notes that the record is "super thick and smaller than today's sized records," later in her review pointing out the fact that its "scratchy record sounds only add to the old-timey charm" (O'Holla, Song Hits of 1939 2014). Her husband, Alex, adds that the record "sounds like someone rubbed sand on it with all its pops and crackling, but somehow that adds to the overall aesthetic of '1939,' if you catch my drift" (O'Holla, Song Hits of 1939 2014). Most of the reviews, other than descriptions of the sleeve art, are spent talking about the records as music, eliding both the technology of the record itself and the equipment on which it is played. The header image for the blog shows a shelf full of records on top of which can barely be seen a Technics turntable and what appears to be a headphones amplifier; there is also a speaker alongside the shelf. None of this equipment gets remarked on in the blog, however.

This elision of the sound-making equipment might be dismissed, as some critics of the blog have done, as conforming to the gender stereotype of the disinterested wife in relation to her husband's technical expertise, but as Jonathan Sterne points out, such an elision of the technology has always been an essential aspect of understanding sound recordings to be "authentic" objects to begin with. Sterne talks about "the ideal form of mediation as a vanishing mediator—where the medium produces a perfect symmetry between copy and original and, thereby, erases itself" (Sterne 2003). The records of *My Husband's Stupid Record Collection* authenticate themselves as "music" rather than as recordings. Whether or not it is warranted, however, the reaction of the blogosphere to O'Holla's blog raises at least a couple of interesting points. First, one is tempted to say so much for the ways in which cyberspace and virtual reality might be said to have the potential to erase social and cultural boundaries, including boundaries of gender. And, second, at this point in the history of recorded music, O'Holla's expressed lack of knowledge about vinyl technology—for instance not knowing what an inner sleeve or a gatefold is—certainly does not need to be stereotyped as "feminine." Many young people of both genders have been raised in an era in which vinyl records were not even on their cultural radar. Some of those young people of both genders are adopting vinyl technology in ever-greater numbers, and O'Holla's blog might be seen as modeling a learning process in relation to vinyl to which they can relate. Furthermore, the distinction between O'Holla being an outsider to the insider experts in the field is somewhat unwarranted. The blog does not position her as merely outside; it positions her as someone who will be experiencing listening to these records and learning from her experience. From the point of view of a reader, O'Holla is an insider to the experiences she is describing. As Hunter Oatman-Stanford notes in his defense of the blog, this project "highlights an undervalued aspect of a decades-long collecting habit: objects like records serve as a proxy for their owners' personal history and offer the opportunity to share these past experiences with others" (Oatman-Stanford 2014). Blogging allows access to non-experts to write about their interests in new and perhaps unfamiliar ways, perhaps raising some defensiveness among the professionals at challenges presented to "official" discourse.

Significantly, O'Holla's very first blog post poses the question "What Is This?" This question resonates in at least three ways that are significant to the present discussion. First, she is asking the question that the blog that follows seeks to answer, "What is a vinyl record?" in terms of why it is worth collecting. Second, she is asking the question, "What is this blog?" in terms of what she is attempting to accomplish by writing it. Third, she is asking the question, "What is a blog?" a question that all blogs seek to answer. Blog formation and identity formation coincide in an attempt to come to terms with the human aspects of vinyl technology. Opie worries that virtual realities and cyberspace "are resoundingly effective in replacing 'natural' encounters with the outside world, yet they present a huge opportunity, filled with unimaginable potential—to find in the virtual world a new way of living, of ordering one's world, even as the old ways are being swept away by uncertainty and placelessness" (2008, 37). What is remarkable, perhaps, given the dire predictions of the potential for virtuality and cyberspace to spin us off into all sorts of wild directions, is the fact that cyber culture has to a great extent stayed pretty close to the ground, if "grounded in physical reality" is meant by "the ground."

Perhaps due to the sheer indefinability and shapelessness of cyberspace, it has to some degree become grounded in the more human-sized aspects of social life and culture. What is one to make of the existence of Etsy, a community based on buying and selling handicrafts? Etsy defines itself as "a marketplace where people around the world connect, both online and offline, to make, sell and buy unique goods" (About Etsy 2015). And, here is how the site defines its mission:

> The heart and soul of Etsy is our global community: the creative entrepreneurs who use Etsy to sell what they make or curate, the shoppers looking for things they can't find anywhere else, the manufacturers who partner with Etsy sellers to help them grow, and the Etsy employees who maintain and nurture our marketplace. (About Etsy 2015)

Among the handicrafts, Etsy also offers an extensive selection of gramophone-related merchandise for sale, including many antiques. Social media builds communities, on both the micro and the macro levels, both local and global communities, based on common interests such as collecting vinyl records, offering possibilities of transgression and transformation to be sure, but also confirming actual lived experience and affirming interests that may have been seen as "anachronistic" or archaic at other times. Furthermore, is it not interesting that in the age of the "selfie," the online users are posting pictures and videos not only of themselves, but also of themselves in relation to the things in which they are interested?

The slogan for YouTube used to be "Broadcast Yourself," a sentiment that could apply to the Internet and social media in general in that it provides unprecedented opportunities for individuals to publish their identities, but the question of how they construct the identities they broadcast remains. Social media are certainly challenging boundaries between self and other, personal and private, but they are also being reestablished in new and different ways. For every over-sharer with no sense of privacy there are others who are using their freedom in different and more creative ways that challenge the norms of established discourse. There exists, for example, on Facebook, an interest group built around collecting 78 records, Edison cylinders, and the machines to play them on. People gather there to talk about their latest acquisitions and to ask and answer questions about buying and restoring antique gramophones and Edison players, and they also post pictures and videos, most of which elicit extended and informed discussion about the early history of recorded music. In one case, a member, a young man of an age at which he might have been more expected to be interested in streaming hip-hop music than in antique shellac records and Victrolas, posted pictures of himself giving a demonstration at a local event in which he was even dressed in period costume. YouTube features hundreds of videos in which people share recordings of 78s and cylinder recordings, including one user who features demonstrations of all of the many antique record players, jukeboxes, and music boxes that he has lovingly restored. He also shares his extensive knowledge of how these antique technologies operate and how to repair them, thereby creating an archive that can be used by future generations of restorers. "These days," Sherry Turkle says,

> cultural norms are rapidly shifting. We used to equate growing up with the ability to function independently. These days, always-on connection leads us to consider the virtues of a more collaborative self. All questions about autonomy look different if, on a daily basis, we are together even when we are alone. (2012, 169)

Turkle worries, perhaps with good reason, that at the same time social media can be used to experiment with identity, it might be "harder to leave the past behind, because the Internet is forever" (2012, 169). A perhaps more interesting question might be what history would look like today if human beings had always been capable of leaving behind such an archive of the stuff of identity formation. Turkle makes the observation that young people in the age of social media "try to conjure a future different from the one they see coming by building on a past they never knew" (2012, 266). Part of that process might just include engagement with a technology well over one hundred years old, the technology of their parents and their grandparents, and in the process of trying to conjure this imagined future, perhaps the future will be made, culturally speaking. Lisa Gitleman suggests that "[h]istorians and critics must be prepared to explain the intensity of modern, mass-cultural experiences as well as their extensive range and appeal" (2008, 84). Part of her point is to argue that in their actions and interactions with technologies, people play a vital role in shaping the outcomes and understandings of the meanings of technologies as much as "corporate strategizing" or technological development. In particular, societal norms and understandings about the mediation of technology in cultural life also involve "normative constructions of difference, whether gender, racial, or other forms" (2008, 84). Rediscovery of vinyl and analog technologies against the digital field of social media alters the ways both analog technologies and social media are understood to be constructed.

Chapter 2 introduced Jason, a vinyl Beatles aficionado who made a YouTube video about his acquisition of a first pressing of *Please Please Me*. Jason is part of an online community that has a presence on both Facebook and YouTube, called the YouTube Vinyl Community (VC). As of this writing, the membership for the Facebook interest group stands at 6,011, while a search of Youtube yields 57,900 video results. The videos mostly consist of members showing off new vinyl purchases or discussing their vinyl collections. Members also offer tours of their listening rooms that include discussions of the equipment on which they play their records. One exception is a 2013 "yearbook" video that includes a montage of pictures that various members of the community submitted of their record shelves (punktexas 2013). Users comment on and discuss each other's videos and subscribe to each other's channels. They also have the option of sharing their videos on Facebook, and many of them do, where they might also solicit subscribers to their YouTube channels. Users might also use contests in which they award prizes in order to solicit new subscribers to their YouTube channels. Some videos appear to have professional looking production values with titles and theme music, while some appear to be no more than a person talking in front of a webcam. Because of the "low barriers to entry" into the VC, Clark Stooksbury notes that "the quality of the videos varies from rank amateur to almost professional." (Stooksbury 2013). Some members post pictures on Facebook but not videos, and some do not post at all, preferring to see and comment on what others are posting.[2]

Stooksbury points to one member as having "the most compelling new channel of 2013," Chinasc12 run by a member named Sydney, who he says,

> can be seen discussing Dylan, Patti Smith, the Velvet Underground, Bikini Kill, and the finer points of Miles Davis and Gil Evans' collaborations in mono, and she occasionally uploads her own covers of songs, such as the Smiths' "I Won't Share You." Sydney not only discusses music from the last six decades with a level of knowledge that belies her brief 14 years on the planet; she also frequently places music in the context of her own life, explaining how different artists inspire her own creative pursuits and match her own personality (Stooksbury 2013)

The Vinyl Community includes members of all ages, genders, nationalities, and ethnicities, among other possible distinctions, the one thing uniting them being an interest in analog records. This openness to crossing social boundaries perhaps makes the VC much more likely to include female members than other analog communities on social media, particularly the ones that relate to equipment and audiophilia. Noting that "the VC is remarkably well-behaved for an online community, Stooksbury quotes member Rob Clark as saying, "Being part of the VC on YouTube has broadened my musical horizons, strengthened my passion for music and cost me a lot of money" (Stooksbury 2013). Like other members of the VC, Clark has made many friends and contacts through his membership in the community. In fact, as of this writing, he was completing a trip to the U.K. with his family in which he met with other members of the VC face to face, and even crate dug with them, all of which he documented on his Facebook feed.

In an ethnography conducted for this book, members of the VC discussed their interests in music and what it means for them to share-broadcast those interests on social media. While they report listening to music on a wide variety of formats, the one thing that ties them all together is their interest in vinyl records. In fact, the responders all reported vinyl as their preferred format, listing reasons such as, "The warmth and the deep natural sound is so intoxicating to my ears. I choose vinyl over any other format 95% of my listening time. I also like the fact that with records you "MUST" sit down and listen to them,"[3] "When I returned to vinyl and started collecting again, my huge love for music returned. There is something about the imperfections and the ability to see and touch and not just hear that seems to make a connection with me,"[4] "Larger tactile format. Takes an effort/intent to actually listen to something, so you pay more attention,"[5] and "One of my uncles had a record collection that I loved immensely and it was him, without even noticing, that made me love vinyl records the way I do."[6] The responses highlight both the influence of other people and the physical aspect of vinyl records as multisensory stimulants. While some members of the VC limit themselves to particular genres of music, for many the format on which they listen to music is much more limited than the genres they listen to, which can be much more adventurous. They might also buy records online, but most respondents report a preference for brick and mortar stores.

When members of the YouTube Vinyl Community are asked what belonging to the community means to them, all of the responses highlight the human connections that they make as a result of such belonging. One member sums these connections up as "Sharing and the community of like minds." He further states, "I have an addiction/passion for vinyl that is not all that common in the grand scheme of things. There's [sic] only a handful of collectors I know of locally, and they are not really friends with common likes."[7] Another member elaborates on that idea, saying,

> I would have to say the most important thing to me is the sense of community caring I feel after I share my records I am spinning. Sometimes you feel like your [sic] the only one that listens to a certain band/artist and when you post it you are just amazed how many more people love that record or artist/band. It reminds you how much music is a universal language and can bring people together![8]

Members seek out others who share their perhaps esoteric tastes, but, just as importantly, they seek validation from other members, engendering a sense of belonging. In fact, at least a double form of validation is taking place, since identifying oneself with vinyl records, or the desire to do so, validates one's decision to join the community in the first place. Identifying oneself and identifying with others are the prime motivating factors for belonging to the Vinyl Community. Ferredey says, "the listing of 'what I think is cool' (Miller 1995) is an essential factor in constructing online identities" (Ferreday 2009, 108). Which is all well and good, as long as others are ready and willing to validate this notion of "cool."

And, of course, Hesmondhalgh's note of caution should also be sounded here as well; he makes the point that "just as self-realization is a highly ambivalent pursuit in modern life, so too are community and sociability" (2013, 47). Andrew Feenberg and Maria Bakardjieva also point out that "[w]hether the Internet contributes to community or undermines it remains an open question" (2004, 2). Sharing tastes also contains an element of status seeking and competitive individualism, and the communities of taste thus created, which is quite easy to discern whenever a member posts a mention or a picture of a Crosley record player as part of a status update. Crosley manufactures inexpensive record players that are often readily available in retail outlets, even stores such as Barnes and Noble that are not generally known for selling electronics (Crosley Cruiser Review 2015). Their price and their ease of availability makes Crosley players a sort of gateway into vinyl listening and collecting for new users often unfamiliar with analog technologies. The fact that they are so cheaply made opens them up for ridicule from more experienced vinyl users. A post featuring a Crosley record player will therefore most of the time invite a barrage of invective and abuse, the friendliest of which will "helpfully" point out that since Crosley players are so cheaply made they probably will destroy the poster's record collection. Also, the sharing of one's tastes in an online forum also carries the risk of ridicule as well as the possibility of validation. The YouTube Vinyl Community on Facebook seems to anticipate both possibilities in its published rules, stating,

> This forum is supposed to be something fun. Not a debate forum, not a political forum, not a forum for airing grievances against other people in the community. Sometimes you will see records posted by artists that you don't like—when that happens, please scroll by—because sometimes you will post a record by an artist that someone else won't like. This is supposed to be about records and music. (YouTube VC n.d.)

To be fair, among the negative and the judgmental comments, many positive and supportive comments will also be found, some even taking the critical posters to task for giving the discussion a negative turn. Administrators even eject members who consistently post negative or demeaning comments. Still, the risks are as much a part of the process as the possibilities. Furthermore, implied in the avowal that one technology is superior is the inference that other technologies are inferior, so there is no escaping critical judgment, however positive one intends the community to be. However, Feenberg and Bakardjieva discuss other theorists' skepticism about what they call "a 'democratic rationalization' of the Internet" (2004, 18). While it might be worth noting that democracy, however rational, is not without its conflicts, also worth noting is their caution that "the social shaping of the Internet is still in process. The technology and its social institution have not yet reached the point of stabilization" (Feenberg and Bakardjieva 2004, 18).

The question remains, however, how an interest in vinyl records authorizes members to create videos about their collections and collecting practices. Baym compares the online communities created by social media to "the concept of the 'speech community,' which foregrounds shared communication practices" (2013, 74). She also makes the related points that these communication practices contribute to the formation of group norms for behavior and interaction (Baym 2013, 76), and that communities tend to be based on "the supportive exchange of resources" (Baym 2013, 78). All of the YouTube Vinyl Community respondents highlight their ability to communicate to other members about their passion as one of their main motivating factors for belonging to the group. As one member says, "I LOVE to share the music I play, and I love the interaction I often get. I have nobody to listen to music with locally, so the community and friendships I am making there will certainly be life long."[9] Another member elaborates in more specific terms; "I tend to express appreciation for the posted recordings that I know, and also comment on posts that are from people with technical problems where I may be of some assistance."[10] There are three specific ways members of the Vinyl Community that highlight the senses of community that Baym appears to be talking about, all of which tend to highlight both the similarities and the differences between online and face-to-face communities. First, the phenomenon of "vibing" tends to highlight the emotional investments between members as it validates and authorizes common tastes. The verb "vibe," according to the *Oxford English Dictionary*, can mean several things, among them, "To transmit or express (a feeling, attitude, etc.) to others in the form of intuitive signals," "To excite, motivate; to send a positive feeling," "To be inspired by or get a positive feeling from a piece of music or (fellow) performer," and "To interact (well), get along, 'click,'"[11] all of which could apply to what members are doing when they post pictures or videos of themselves listening to the same record(s) as other members at roughly the same time. The *Urban Dictionary* perhaps puts it more succinctly when is defines "vibing" as "1. Hanging, doing nothing, chilling" and 2. "listening to music" (Miller and Perez 2008). Members tend to "vibe with" other members in their posts, highlighting the double act of validation previously discussed, which one member expressed in response to a question about members whose videos he follows; "I watch new ones on occasion to see if there are other collectors who like the same genres. I tend to only follow the ones that have similar tastes to mine, as I like seeing new things related to what I enjoy personally."[12] Keeping in mind Hesmondhalgh's caution about the competitive aspects of the display of tastes, the act of vibing also allows members to share their tastes and at the same time validate and be validated by the tastes of other members.

The second type of communication involves the sharing of resources Baym talked about; it is called "VCLT," or "Vinyl Community Love Train." Members send albums to each other that they think the others will enjoy, and the members who receive the VCLT post either videos or photos, many of which feature the members opening the package on camera and sharing their surprise with their viewers. Baym describes this phenomenon as "bridging capitol," which "is exchanged between people who differ from one another and do not share strong relationships" (Baym 2013, 78). Baym notes "online groups provide bridging capital, exchanged in relationships that are highly specialized, yet it is also common to find members of online communities and social networks providing one another with the sort of emotional support often found in close relationships" (Baym 2013, 78-79). She refers to Cutrona and Russell to call this type of exchange "tangible aid" as opposed to other types of emotional and informational support that online communities offer (Baym 2013, 81). In the case of VCLT, it appears that just as much as it bonds group members together and validates common tastes, it is also a way to expanding other group members' horizons based more or less on common interests. Baym makes the point that "[b]eing a skilled provider also increases people's status and prestige within online groups" (Baym 2013, 81). It can also have commercial implications, since owners of independent record labels—who wish to publicize products they have for sale on the channels of those members to whom they send the VCLT—send some of them.

The third type of communication this discussion wishes to highlight is the meet-up and crate digging among VC members. Baym notes that many people form "strong ties" on social media, ties that she says, "are voluntary, mutually reciprocal, supportive of the partners' needs, and they create long-term contact" (Baym 2013, 114). Some members of the VC establish these types of long-term relationships with other members. Some members even travel either short or long distances in order to meet other members. In cases such as these, two or more members of the VC will post a picture or a video of the meet-up, and in some cases these meet-ups even involve the exchange of VCLT. In other cases, members meet to go to record stores, crate-digging as a means of building social bonds. These members also post pictures and videos of their meet-ups, in which they also sometimes display their purchases. These types of face-to-face meetings and the communication they engender are not exclusive to the VC, but one thing that can be said for them that is vinyl specific is that because of the scarcity of brick-and-mortar record stores, particularly in more rural areas, vinyl users are used to travelling greater distances specifically to indulge their passions, crate-digging for vinyl where they can find it, which could be said to sanction the kinds of effort that such travel requires.

Hesmondhalgh asserts his own assent to Lawrence Kramer's notion that "music of all kinds embodies the drive for attachment" (2013, 79). Kramer himself says that music, "works, it grips or grasps us, almost with the electricity of touch, resonant, perhaps, with the primary experiences of bonding that tie us to each other and the world" (2007, 33). Whether or not the types of bonding over music provided by the Internet and social media occasion a positive good is a question that this chapter and this book cannot in the final analysis answer. After noting that, "Such thinking persists in the hope that new digital technologies can democratize culture, by leading to the breakdown of capitalist property relations, or by empowering 'consumers' to become 'producers,'" Hesmondhalgh discusses Bourdieu's difficulty with the notion with the emancipatory potential of shared aesthetic experience (2013, 114-115) According to Hesmondhalgh, in his study *La Distinction*,

> Bourdieu showed how tastes were linked to particular sets of dispositions associated with particular social classes and how, in their cultural practices, dominant classes accrued "cultural capital" which they could, in effect, use to maintain and leverage their superiority. The cultural practices of the working class were despised, dismissed, and misunderstood. (2013, 115)

Hesmondhalgh himself concludes that despite our difficulties in conceiving of a common aesthetic language, "many people find in musical experience powerful and moving evocations of collectivity and solidarity" (2013, 122). However, he also cautions that "we should be wary about overloading aesthetic deliberation with hopes of a historical role for aesthetic experience in modern life" (Hesmondhalgh 2013, 122). Whatever its emancipatory possibilities, however, the information age *has* enabled what might be identified as perhaps counter-discursive aesthetic practices to emerge and to flourish, and, for good or for ill, the return of analog music and vinyl records are among them. YouTube and other social media are to a great degree changing notions of "content" and "expertise" in ways that were unimaginable prior to the information age. Michael Strangelove says that YouTube, "represents a transformation in the structure of our media-saturated culture. This transformation is both simple and profound in its consequences. This is the transformation of *who is saying what to whom*" (2010, 16). Since the transformation is still taking place, the answer to Strangelove's question is still very much open.

Communities such as VC, and the personal relationships they engender, to a great degree take the vastness and incomprehensibility of cyberspace and shrink it down to a more human size. Ultimately, what gives the VC and its members meaning is the shared interest in vinyl records. The videos they generate about their records, which may or may not involve playing the records themselves, contribute to a cyber archive that is validated by the shared perception that vinyl provides a less impersonal, more "human" form of interaction, one of the reasons often given for vinyl's comeback in the first place. Or, perhaps the act of showing one's collection online might better be seen as the assertion of one's humanity in the face of the perception that culture is rapidly dematerializing. Jonathan Sterne discusses the impulse toward archive building as part of the preservative function of recording technology (2003, 325–333). This chapter and this book will conclude that in the digital realm, efforts such as those of the YouTube Vinyl Community, the videos they make and the pictures and stories about vinyl records they post online to at least some degree participate in a similar archival project, an archive of individual identities in relation to common interests in vinyl recordings. Such impulses place these efforts on a similar plain with fan fiction, which Abigail Derecho borrows Jacques Derrida's term, "archontic," which designates a type of archive that is "ever expanding and never completely closed" (2006, 57). Archontic texts "build on a previously existing text" in a way that "allows, or even invites, writers to enter it, select specific items they find useful, make new artifacts using those found objects, and deposit the newly made work back into the source text's archive" (Derecho 2006, 60). While the bricoleurs of the blogs and other social media about vinyl records are not necessarily doing exactly the same thing as fan fiction, they do share a similar intertextual impulse in the ways that users interact with the recordings they are writing about, thus altering the ways in which one might think about both the record and the individual, blurring boundaries at the same time they validate and reinforce each other.

One question this ever-expanding archive of profiles, posts, and responses might raise in the case of vinyl records is what happens to this accumulation of data. Where does it all lead? So far, it has led to the expansion of interest in vinyl records, to the point at least at which commentators are calling it no longer a niche market. What happens when this social and cultural phenomenon reaches critical mass, and what would critical mass even look like? What happens when talking about vinyl is no longer "anachronistic?" What becomes of these identities so carefully constructed around vinyl's outsider status? Evan Eisenberg notes about digital technology, "unlike text or images, music doesn't lend itself to being scanned or 'surfed' (except, perhaps, by musicians with particular questions in mind). Music happens in real time. Despite all the talk about real time in cyber circles, few cybernauts have the patience for *real* real time" (2005, 215). Perhaps the fact that analog music cannot be surfed is the real reason for the unexpected return to vinyl and analog recordings, but even if it is, the most fascinating prospect about this return could just be that the cybernauts are turning out to be the great evangelists of the comeback.

NOTES

1. Felix Atagong, personal email to the author, June 28, 2015.
2. In the interest of disclosure, I should say at this point that I am a member of the Facebook YouTube Vinyl Community group and that I do post photos of the records I am currently playing periodically.
3. Michael Turner, response to questionnaire, January 9, 2015.
4. Jonathan Henschel, response to questionnaire, January 26, 2015.
5. Alex McCubbin, response to questionnaire, January 6, 2015.
6. Álvaro Martín Gómez Acevedo, response to questionnaire, January 2, 2015.
7. Alex McCubbin, response to questionnaire, January 6, 2015.
8. Michael Turner, response to questionnaire, January 9, 2015.
9. Jonathan Henschel, response to questionnaire, January 26, 2015.
10. Brad Wood, response to questionnaire, January 2, 2015.
11. "vibe, v.". OED Online. June 2015. Oxford University Press. http://www.oed.com/view/Entry/247856?result=2&rskey=UqIrmw& (accessed June 30, 2015).
12. Alex McCubbin, response to questionnaire, January 6, 2015.

The Run-Out Groove

Among many others, one of the more admirable things about Evan Eisenberg's seminal work on recorded sound, *The Recording Angel*, was the sheer foolhardy audacity it took to conclude with a science fiction story that attempted to predict the future of the consumption of recorded music. The answer to the question of whether or not the technology of vinyl records will ever regain its footing as the dominant format for the consumption of music, while doubtful at best, is nonetheless interesting to consider. Eisenberg makes the point in his introduction that although his conclusions are speculative, they are in fact based on technologies that were currently under development at the time the book was published (2004) (2005, viii). Eisenberg imagines a world in which human beings are becoming fused with sound technologies, to the degree where they listen to recorded music from within their own bodies in various ways. One character, Wack, has had his ears placed surgically on the inside, saying that it "Makes things sound cool, too. More bass." (Eisenberg 2005, 218). Eisenberg also imagines a world in which fans can either join or become their favorite performers on stage in a virtual reality version of Rock Band, and in which people can stream any type of music they like from inside of their own bodies using "The Celestial Jukebox" (2005, 221). He imagines a world in which users can make music with their bodies through synthetic means by wearing a special pendent, concluding, "[t]he line between 'musicians' and 'listeners' seems to have been erased" (Eisenberg 2005, 224). Music matches itself to people's moods and both stimulates and satisfies their desires. Furthermore, in this imagined future, music is still to some degree controlled and serves the corporate interests of the music industry, even if it is somewhat vague about how this all works.

To conclude, this book will not be making any predictions about the future, however. It will make a couple of observations based on Eisenberg's speculations, though. First, of all of Eisenberg's wildest imaginings, however based on real technologies then in development they might have been, the one thing he did not bank on was the late return of the vinyl record. For him, as for Mark Katz in *Capturing Sound*, vinyl records are pretty much a dead issue. To be sure, one of the things that is remarkable about vinyl's comeback is its sheer unpredictability. It seems to be growing on its own, nurtured by the Internet, outside of the eye of culture writers and the music industry alike. Who can say whether this phenomenon will last or continue to grow? Nobody really predicted that it would happen in the first place. The one thing that can be said is that brick-and-mortar record stores are being established and welcoming customers in the year 2015, twenty-five years after vinyl was declared a dead format and consigned to the back rooms of antique stores and thrift shops. Furthermore, electronics manufacturers are making and selling turntables again, and new analog, yes analog, technologies are being developed. Factories for vinyl records are behind their pressing schedules, and new factories are being equipped, while old vinyl manufacturing equipment is retooled to meet current needs. None of this was on the radar in 2004 when Eisenberg was writing, and it had only just gotten going at the time Katz was.

The one thing that Eisenberg's prediction got right, no matter how sound technology develops in the coming years, is the connection he makes between human beings and their perceptions and the music they consume. If the technology develops to the point where we are making music inside our own heads, it will only be a further iteration of what we are already doing. Regardless of how we choose to listen to music technologically speaking, we are always already making the music we perceive in our own minds, using our own nervous systems. The technology does not make us, as much as the music and the electronics industries would like us to think that it does. We make the music in the ways that we interact with the technology, and we choose the ways we make it in choosing the ways we listen to it, no matter what advertising and popular culture tell us. Controversies and court cases surrounding file sharing and streaming might be about the status of intellectual property in the age of limitless mechanical reproduction, but they are also about who controls the narrative, who gets to say which music is important and to dictate the ways in which it should be listened. Vinyl users are taking back those decisions for themselves and coopting the official narratives that commercialized culture attempts to impose on them. Technology ultimately may or may not be predictable, people and the cultures that they make up are far less so.

Bibliography

"9 Things Audiophiles Hate." *HDtracks*. HDtracks. January 9, 2014. http://www.hdtracks.com/blog/?p=272 (accessed June 3, 2015).

"About Light in the Attic." *Light in the Attic Records*. 2015. http://lightintheattic.net/about (accessed February 26, 2015).

Adorno, Theodor W. "The Culture Industry Reconsidered." In *Dialectic of Enlightenment: Philosophical Fragments*, by Max Horkheimer and Theodor W. Adorno, translated by Edmund Jephcott, 94–136. Stanford: Stanford University Press, 2002.

"All Eyez on Me [Vinyl] Explicit Lyrics, Original recording remastered." *Amazon.com*. May 22, 2001. http://www.amazon.com/All-Eyez-Me-Vinyl 2pac/dp/B00005AQE7/ref=tmm_vnl_title_0?_encoding=UTF8&sr=1-1&qid=1429069521 (accessed April 15, 2015).

Altschuler, Glenn C. *All Shook Up: How Rock 'n' Roll Changed America*. Cary, NC: Oxford University Press, 2003.

Anderson, Tim J. *Making Easy Listening: Material Culture and Postwar American Recording*. Minneapolis, MN: University of Minnesota Press, 2006.

"Apple Launches iTunes Plus." *Apple.com*. May 30, 2007. https://www.apple.com/pr/library/2007/05/30Apple-Launches-iTunes-Plus.html (accessed February 8, 2015).

Atkinson, John. "Stereo and the Soundstage." *Stereophile*. January 4, 2008. http://www.stereophile.com/asweseeit/1286awsi/ (accessed April 29, 2015).

Attali, Jacques. *Noise: The Political Economy of Music*. Translated by Brian Massumi. Minneapolis: University of Minnesota Press, 1985.

Backhouse, Jim. "Jim Backhouse: The Beyonder." *Record Collector*, February 2014: 12.

Bakhtin, M. M. *The Dialogic Imagination: Four Essays*. Edited by Michael Holquist. Translated by Caryl Emerson and Michael Holquist. Austin: University of Texas Press, 1981.

Banash, David. "Nostalgia, Distinction, and Collecting in the Twenty-First Century." In *Contemporary Collecting: Objects, Practices, and the Fate of Things*, edited by Kevin M. Moist and David Banash. Lanham, MD: Scarecrow Press, 2013.

Barker, Hugh, and Yuval Taylor. *Faking It: The Quest for Authenticity in Popular Music*. New York & London: W.W. Norton & Co., 2007.

Bartmanski, Dominick, and Ian Woodward. *Vinyl: The Analog Record in the Digital Age*. London & New York: Bloomsbury, 2015.

Baudrillard, Jean. "Simulacra and Simulations." In *Selected Writings*, by Jean Baudrillard, edited by Mark Poster, 166–184. Stanford, CA: Stanford University Press, 1988.

———. "The System of Collecting." In *The Cultures of Collecting*, edited by John Elsner and Roger Cardinal, 9–24. Cambridge, MA: Harvard University Press, 1994.

Baym, Nancy K. *Personal Connections in the Digital Age.* New York: John Wiley & Sons, 2013.
Bazin, André. *What Is Cinema? Volume 1* Translated by Hugh Gray. Oakland: University of California Press, 1967.
———. *What Is Cinema?, Volume 2.* Translated by Hugh Grant. Berkeley & Los Angeles, CA: University of California Press, 1971, 2005.
Beedle, Ashley. "Ashley Beedle: Dancefloor Polymath." *Record Collector*, 2014: 14.
Belam, Martin. "It Was Twenty Years Ago Today . . . The Beatles CD reissues from 1987." *currybetdotnet.* April 17, 2007. http://www.currybet.net/cbet_blog/2007/04/it-was-twenty-years-ago-todayt.php (accessed June 6, 2015).
Benjamin, Walter. "Unpacking My Library: A Talk about Book Collecting". In *Illuminations*, edited by Hannah Arendt, translated by Harry Zohn. New York: Schocken Books, 1968.
———. "The Work of Art in the Age of Mechanical Reproduction." In *Illuminations*, edited by Hannah Arendt, translated by Harry Zohn, 217–251. New York: Schocken, 1968.
Bennett, Tony. "Putting Policy into Cultural Studies." *Cultural Studies*, edited by Lawrence Grossberg, Cary Nelson, and Paula Treichler, 23–37. New York: Routledge, 1992.
Berger, John. *Ways of Seeing.* Harmondsworth: Penguin Books, 1973.
Berker, Thomas, Maren Hartmenn, Yves Punie, and Katie Ward. "Introduction." In *Domestication of Media and Technology*, by Thomas Berker, Maren Hartmenn, Yves Punie and Katie Ward. Maidenhead, Berkshire: Open University Press, 2006.
Biancholli, Louis, and Lester H. Bogen. *Understanding High Fidelity: A Guide to Hi-Fi Home Music Systems.* New York: David Bogen Company, Inc., 1953.
Bilton, Nick. "Apple Teases Customers with iTunes Announcement." The *New York Times.* November 15, 2010. http://bits.blogs.nytimes.com/2010/11/15/apple-teases-customers-with-itunes-announcement/ (accessed June 6, 2015).
Blender, Morton. *High Fidelity*, March 1954: 21–23.
Borgmann, Albert. 2009. *Technology and the Character of Modern Life: A Philosophical Inquiry.* revised, Chicago: University of Chicago Press.
boyd, danah. "Participating in the Always-On Lifestyle." In *The Social Media Reader*, edited by Michael Mandiberg, 71–76. New York & London: New York University Press, 2012.
Bratich, Jack Z., Jeremy Packer, and Cameron McCarthy. "Governing the Present." In *Foucault, Cultural Studies, and Governmentality*, edited by Jack Z. Bratich, Jeremy Packer, and Cameron McCarthy, 3–22. Albany: State University of New York Press, 2003.
Brittan, Gordon G., Jr., "Technology and Nostalgia." In *Technology and the Good Life?*, by Eric Higgs, Andrew Light, and David Strong, 70–88. Chicago & London: University of Chicago Press, 2000.
Bürger, Peter. "The Negation of the Anatomy of Art by the Avant-Garde." In *Postmodernism: A Reader*, edited by Thomas Docherty, 237–243. New York, Oxford: Columbia University Press, 1993.
Burke, Kenneth. *A Grammar of Motives.* Berkeley: University of California Press, 1945.
———. *Language as Symobolic Action: Essays on Life, Literature, Method.* Berkeley, CA: University of California Press, 1966.
Bush, John. "The SMiLE Sessions." *All Music.* 2011. http://www.allmusic.com/album/the-smile-sessions-mw0002215480 (accessed March 12, 2015).
———. "Various Artists: Anthology of American Folk Music, Vol. 1–3." *All Music.* 1997. http://www.allmusic.com/album/anthology-of-american-folk-music-vol-1-3-mw0000028056 (accessed March 12, 2015).
Calamar, Gary, and Phil Gallo. *Record Store Days.* New York: Sterling, 2009.
Campbell, Gerry. "Group Announcement. Posting Guidelines Additions. Please Read Immediately." *Facebook.* April 26, 2015. https://www.facebook.com/groups/640334122671073/ (accessed May 4, 2015).
Campbell, Neil. "Neil Campbell: Star Guzzler." *Record Collector*, September 2014: 22.
Carnes, Richard. "Discogs: Vinyl Revolution." *Resident Advisor.* March 26, 2010. http://www.residentadvisor.net/feature.aspx?1166 (accessed June 19, 2015).

Caro, Mark. "Going Mono a Mano with the Beatles." The *Chicago Tribune*. September 30, 2014. http://www.chicagotribune.com/entertainment/music/chi-beatles-mono-vinyl-box-review-20140930-column.html#page=1 (accessed June 13, 2015).
———. "The Ultimate Beatles Sound Test." The *Chicago Tribune*. November 12, 2012. http://articles.chicagotribune.com/2012-11-12/entertainment/ct-ent-1113-beatles-vinyl-20121112_1_beatles-nostalgia-beatles-catalog-beatles-lps (accessed June 14, 2015).
Carr, Nicholas. *The Shallows: What the Internet Is Doing to Our Brains*. New York: W.W. Norton, 2010.
"Caruso 2000." *Amazon.com*. February 8, 2000. http://www.amazon.com/Caruso-2000-Enrico/dp/B000031WTO (accessed June 7, 2015).
Caulfield, Keith. "Beatles Stir Mono Mania." *Billboard*, September 27, 2014: 69.
Cavell, Stanley. "The World as Things: Collecting Thoughts on Collecting." In *Contemporary Collecting: Objects, Practices, and the Fate of Things*, edited by Kevin M. Moist and David Banash. Lanham: The Scarecrow Press, 2013.
Cerra, Allison, and Christina James. *Identity Shift : Where Identity Meets Technology in the Networked-Community Age*. Hoboken, NJ: John Willey & Sons, 2011.
Christgau, Robert. "Anthology of American Folk Music: Smithsonian Folkways." *Robert Christgau: Dean of American Rock Critics*. October 1997. http://www.robertchristgau.com/xg/cdrev/smithson-cut.php (accessed March 12, 2015).
Christman, Ed. "Beach Boys Engineer Mark Linett Talks 'Smile' Release." *Billboard*. March 11, 2011. http://www.billboard.com/articles/news/472565/beach-boys-engineer-mark-linett-talks-smile-release (accessed March 11, 2015).
———. "Beatles Being Paid Directly by iTunes in Deal." *Reuters*. January 5, 2011. http://www.reuters.com/article/2011/01/06/us-beatles-itunes-idUSTRE7050IC20110106 (accessed June 13, 2015).
———. "Vinyl Solution? One Physical Format Is Doing Better this Year." *Billboard*, April 28, 2007: 11.
———. "Record Store Day Sets Twelve-Year Sales High for Indies, 'Certainly One for the Books.'" *Billboard*, April 23, 2015. http://www.billboard.com/articles/news/6539051/record-store-day-sets-records-twelve-year-sales-high-indies (accessed April 30, 2015).
Classic Cover Ups. 2013. http://www.anorakscorner.com/CoverUps.html (accessed March 20, 2015).
"Columbia House Company History." *Funding Universe*. Accessed January 1, 2016. http://www.fundinguniverse.com/company-histories/columbia-house-company-history/.
Constantinides, John. "The Sound System: Contributions to Jamaican Music and the Montreal Dancehall scene." *Dread Library: Roots Rock Reggae*. 2002. http://debate.uvm.edu/dreadlibrary/constantinides2004.htm (accessed March 20, 2015).
Cook, Deborah. "The Sundered Totality of System and Lifeworld." *Historical Materialism* 13, no. 4 (2005): 55–78.
"Crosley Cruiser Review." *What Hi Fi?* October 12. Accessed December 31, 2015. http://www.whathifi.com/crosley/cruiser/review.
Currie, Mark. "The Hectic Quest for Prelapsarian Man: the Adamic Myth in Late Auden." *Critical Survey* (Berghahn Books) 6, no. 3 (1994): 355–360.
Currin, Grayson. "Various Artists The Rise and Fall of Paramount Records, Volume One (1917–1932) Third Man; 2013." *Pitchfork*. Pitchfork Media Inc. November 22, 2013. http://pitchfork.com/reviews/albums/18703-the-rise-fall-of-paramount-records-volume-one-1917-1932/ (accessed March 27, 2015).
Dankosky, John. "Why Vinyl Sounds Better than CD, or Not." *NPR*. February 12, 2012. http://www.npr.org/2012/02/10/146697658/why-vinyl-sounds-better-than-cd-or-not (accessed March 26, 2013).
———. "Why Vinyl Sounds Better Than CD, Or Not." *NPR* February 12, 2012. http://www.npr.org/2012/02/10/146697658/why-vinyl-sounds-better-than-cd-or-not (accessed February 4, 2015).
Davis, John. "Going Analog: Vinylphiles and the Consumption of the 'Obsolete." In *Residual Media*, edited by Charles R. Acland, 222–236. Minneapolis: University of Minnesota, 2007.

Davis, Jonathan. "Questioning 'The Work of Art in the Age of Mechanical Reproduction': A Stroll around the Louvre after Reading Benjamin." *Contemporary Aesthetics.* January 11, 2008. http://www.contempaesthetics.org/newvolume/pages/article.php?articleID=493 (accessed June 10, 2015).

De Certeau, Michel. *The Practice of Everyday Life.* Translated by Steven F. Rendall. Berkeley: University of California Press, 2002.

Dean, Mitchell. *Governmentality: Power and Rule in Modern Society.* London, Thousand Oaks, New Delhi: Sage Publications, 1999.

DeChaine, Robert D. "Memory, Desire, and the 'Good Collector' in PEZhead Culture (2013-05-09). Contemporary Collecting: Objects, Practices, and the Fate of Things (Kindle Locations 1471–1472). Scarecrow Press. Kindle Edition." In *Contemporary Collecting: Objects, Practices, and the Fate of Things*, edited by Kevin M. Moist and David Banash. Lanham, Toronto & Plymouth, UK: The Scarecrow Press, Inc., 2013.

Derecho, Abigail. "Archontic Literature: A Definition, a History, and Several Theories of Fan Fiction." In *Fan Fiction and Fan Communities in the Age of the Internet*, edited by Karen Hellekson and Kristina Busse, 57–74. Jefferson, NC and London: McFarland & Company, Inc., 2006.

Designboom. *The Nipper Saga.* 2000–2010. http://www.designboom.com/history/nipper.html (accessed November 30, 2014).

Dowling, William. *Beatlesongs.* New York: Fireside, 1989.

Dunn, Robert G. *Identifying Consumption : Subjects and Objects in Consumer Society.* Philadelphia: Temple University Press, 2008.

Durrans, Brian. "Behind the Scenes Museums and Selective Criticism ." *Museums, Representation and Cultural Property.* 1992. http://www.prm.ox.ac.uk/Kent/musantob/thobrep2.html (accessed April 27, 2015).

"E. Berliner's Gramophone. Directions for Users of the Seven-Inch American Hand Machine." Washington, DC: The United States Gramophone Co., 1894.

Eder, Bruce. "The Beatles Reel Music." *All Music.* http://www.allmusic.com/album/reel-music-mw0000319111 (accessed June 7, 2015).

Eisenberg, Evan. *The Recording Angel: Music, Records and Culture from Aristotle to Zappa.* 2nd edition. New Haven, CT: Yale University Press, 2005.

Elborough, Travis. *The Vinyl Countdown: The Album from LP to iPod and Back Again.* Brooklyn, NY: Soft Skull Press, 2009.

"Elvis Presley Biography: A Comprehensive History of Elvis Presley's Dynamic Life—See more at: http://biography.elvis.com.au/#sthash.fKpyNoGW.dpuf." *Elvis Australia.* http://biography.elvis.com.au/ (accessed April 22, 2015).

Emerick, Goeff, and Howard Massey. *Here, There and Everywhere: My Life Recording the Music of the Beatles.* New York & London: Gotham Books, 2006.

EMI Records Ltd. "Nipper and His Master's Voice: What Is the Story?" EMI Records Ltd., 1997.

Erlewine, Stephen Thomas. "The Beatles The Capitol Albums, Vol. 1." *All Music.* 2004. http://www.allmusic.com/album/the-capitol-albums-vol-1-mw0000139507.

Fantel, Hans H. "Stereo Record Players: How Five New Intergrated Players, Improved in Concept and Quality, Meet the Challenge of Stereo." *HiFi/Stereo Review*, February 1960: 60–66.

Farberman, Brad. "Bucks Burnett's Very Cool Eight Track Museum Opens This Weekend." *The Village Voice Blogs.* Village Voice LLC. October 12, 2012. http://blogs.villagevoice.com/music/2012/10/bucks_burnetts.php (accessed April 26, 2015).

Farmer, Nick. "Please Please Me—Fixing a Hole." *Record Collector*, May 2012: 74–80.

Feedbands. 2015. *Feedbands Launches Farm to Feed Bands.* September. Accessed December 1, 2015. https://feedbands.com/farm/#.

Feenberg, Andrew, and Maria Bakardjieva. "Consumers or Citizens: The Online Community Debate." In *Community in the Digital Age: Philosophy and Practice*, by Andrew Feenberg and Darin Barney, 1–28. New York: Rowman & Littlefield, 2004.

Feldman, Karen S. "Not Dialectical Enough: On Benjamin, Adorno, and Autonomous Critique." *Philosophy and Rhetoric* 44, no. 4 (2011): 337–362.

Ferreday, Debra. *Online Belongings: Fantasy, Affect and Web Communities*. Oxford: Peter Lang AG, 2009.
Figueroa-Serriera, Heidi J. "Connecting the Selves Computer-Mediated Identification Processes." In *Critical Cyberculture Studies*, translated by David Silver and Adrienne Massanari, 97–106. New York: New York University Press, 2006.
Fingas, Jon. 2015. "Columbia House Hopes You'll Come Back for Vinyl Records." *Engadget.* December 26. Accessed January 1, 2016. http://www.engadget.com/2015/12/26/columbia-house-vinyl-hopes/.
Fitzpatrick, Rob. "The Beginning: From Scallies to Stars." *Q Magazine*, May 2013.
Flanagan, Andrew, and Mark Schneider. *Neil Young Calls Vinyl Comeback a 'Fashion Statement' Amidst High-Quality Audio Quest*. February 3, 2015. http://www.billboard.com/articles/business/6458311/neil-young-calls-vinyl-comeback-a-fashion-statement (accessed February 4, 2015).
Foucault, Michel. *Discipline and Punish: The Birth of the Prison*. Translated by Alan Sheridan. New York: Random House, 1979.
———. "History of Systems of Thought." In *Language, Cunter-Memory, Practice: Selected Essays and Interviews*, by Michel Foucault, edited by Donald Bouchard, translated by Donald Bouchard and Sherry Simon, 199–204. Ithaca, NY: Cornell University Press, 1977.
———. "Intellectuals and Power: A Conversation between Michel Foucault and Gilles Deleuze." In *Language, Counter-Memory, Practice: Selected Essays and Interviews*, edited by Donald F. Bouchard, translated by Donald F. Bouchard and Sherry Simon, 205–217. Ithaca, NY: Cornell University Press, 1977.
———. *Power/Knowledge: Selected Interviews and Other Writings, 1972–1977*. Edited by Colin Gordon. Translated by Colin Gordon, Leo Marshall, John Mepham and Kate Soper. New York: Pantheon Books, 1980.
Foucault, Michel. "Questions of Method." In *The Foucault Effect: Studies In Governmentality with Two Lectures and an Interview with Michel Foucault*, edited by Graham Burchell, Colin Gordon, and Peter Miller, translated by Colin Gordon, 73–86. Chicago: University of Chicago Press, 1991.
Fremer, Michael. "Don't Let This Happen to You." *Analogue Planet.* May 2, 2013. http://www.analogplanet.com/content/dont-let-happen-you (accessed May 31, 2015).
———. *The Turtles 45rpm Box Set: "Happy Un-Together"?* November 4, 2014. http://www.analogplanet.com/content/turtles-45rpm-box-set-happy-un-together (accessed February 4, 2015).
Gadamer, Hans-Georg. *Truth and Method*. Translated by Joel Weinsheimer and Donald G. Marshall. London: Bloomsbury, 1988.
Gallucci, Michael. "Top 10 Beatles Bootleg Albums." *Ultimat Classic Rock.* February 16, 2013. http://ultimateclassicrock.com/beatles-bootleg-albums/ (accessed June 13, 2015).
Garland, Emma. "Does Record Store Day Screw Small Labels?" *The 405.* March 24, 2014. http://www.thefourohfive.com/news/article/does-record-store-day-screw-small-labels-139 (accessed April 30, 2015).
Garnett, Robert. "Too Low to be Low: Art, Pop, and the Sex Pistols." In *Punk Rock: So What? The Cultural Legacy of Punk*, edited by Roger Sabin, 17–30. London & New York: Routledge, 1999.
Gilbert, Pat. "The Beatles In Mono Vinyl Box Set." *Mojo* , October 2014: 80.
———. "Back to Mono." *Mojo*, October 2014: 78–82.
Gitelman, Lisa. *Always Already New: Media, History, and the Data of Culture*. Cambridge, MS: The MIT Press, 2008.
———. *Scripts, Grooves, and Writing Machines: Represnting Technology in the Edison Era.* Stanford, CA: Stanford University Press, 1999.
Glasgow, Joshua. "Hi-Fi Aesthetics." *The Journal of Aesthetics and Art Criticism* 65, no. 2 (Spring 2007): 163–174.
Goodrum, Charles, and Helen Dalrymple. *Advertizing In America: The First Two Hundred Years.* New York: Harry N. Abrams, 1990.

Gracyk, Tim. "Music That Americans Loved 100 Years Ago—Tin Pan Alley, Broadway Show Tunes, Ragtime, and Sousa Marches." *Tim's Phonographs and Old Records*. 2006. http://www.gracyk.com/century.shtml#coon (accessed February 19, 2015).

Grajeda, Tony. "'The Sweet Spot:' The Technology of Sound and the Field of Auditorship." In *Living Stereo: Histories and Cultures of Multichannel Sound*, edited by Paul Théberge, Kyle Devine, and Tom Everrett, 37–64. New York & London: Bloomsbury, 2015.

Greenblatt, Stephen. *Renaissance Self-Fashioning: From More to Shakespeare*. Chicago: University of Chicago Press, 2012.

Guglielmo, Connie. "Apple, Capitalizing on British Invasion, Creates Dedicated Channel For the Beatles on iTunes. ." *Forbes*, February 10, 2014: 5.

Gundersen, Edna. "Digital Stores Now Leading Source for Album Sales." *USA Today*. Gannett. January 24, 2013. http://www.usatoday.com/story/life/music/2013/01/23/digital-stores-beat-mass-merchants-in-album-sales/1843627/ (accessed March 25, 2013).

Haley, John H. "Sound Recording Reviews: The Beatles in Mono (LP Edition)." *ARSC Journal*, 2015: 151–154.

Hand, Martin. "A People's Network: Access and the Indefiniteness of Learning." In *Making Digital Cultures: Access, Interactivity, and Authenticity*, Abingdon, Oxon: Ashgate Publishing Group, 2008, 75–99.

Harley, Robert. "An LP Primer: How the LP Works." *The Absolute Sound*, June/July 2007: 35–39.

Hayes, David. " 'Take Those Old Records off the Shelf:' Youth and Music Consumption in the Postmodern Age." *Popular Music and Society* 29, no. 1 (February 2006): 51–68.

Heatley, Michael. "Prog's Never-Netherlands." *Record Collector*, no. 430 (August 2014): 32–38.

Hesmondhalgh, David. *Why Music Matters*. Chichester: Wiley-Blackwell, 2013.

Hess, Amanda. "A Woman Reviews Her Husband's 'Stupid' Record Collection. Is That Sexist, Funny, or Both?" *Slate*. March 18, 2014. http://www.slate.com/blogs/xx_factor/2014/03/18/my_husband_s_stupid_record_collection_sarah_o_holla_reviews_music_nerd_husband.html (accessed June 28, 2015).

Heylin, Clinton. *Bootleg: The Secret History of the Other Recording Industry*. New York: Macmillan, 1996.

Hickey, Matthew. 2015. "The Best Vinyl Subscription Services For Record Collectors." *Turntable Kitchen*. July 23. Accessed January 1, 2016. http://www.turntablekitchen.com/2015/07/the-best-vinyl-subscription-services-for-record-collectors/.

High Fidelity. "Noted With Interest." March 1954: 11.

High Fidelity. "Quality is a Word Often Loosely Used." March 1954.

High Fidelity. "The New Harmon Kardon High Fidelity AM-FM Tuner." March 1954: 4.

Hinson, Peter. "Paul Weller Is Done with Record Store Day." *The 405*. April 22, 2014. http://www.thefourohfive.com/news/article/paul-weller-is-done-with-record-store-day-139 (accessed April 30, 2015).

Hoffman, Steve. "Frequently Asked Questions About the Sound of CDs." *Steve Hoffman TV*. May 3, 2003. http://www.stevehoffman.tv/dhinterviews/HoffSound.htm (accessed March 26, 2013).

Hogan, Marc. "Did Vinyl Really Die in the '90s? Well, Sort Of.…" *Spin*. May 16, 2014. http://www.spin.com/2014/05/did-vinyl-really-die-in-the-90s-death-resurgence-sales/ (accessed June 15, 2015).

———. "The Beatles Finally Reissue Remastered LPs on Vinyl." *Spin* magazine. September 27, 2012. http://www.spin.com/2012/09/the-beatles-remastered-lp-vinyl-reissue/ (accessed June 6, 2015).

Holmes, Thom. "Shellac." In *The Routledge Guide to Music Technology*, by Thom Holmes, 278. London & New York: Routledge, 2013.

Holt, J. Gordon. "It's the Real Thing!" *Stereophile.com*. October 22, 2014. http://www.stereophile.com/content/its-real-thing (accessed February 5, 2015).

Homer, Sean. 2005. *Jacques Lacan*. London & New York: Routledge.

Humphries, Patrick. "The Ship Finally Comes In." *Record Collector*, January 2015: 72–77.

Bibliography

"John Beaulieu's Facebook page." *Facebook.* Facebook, Inc. April 19, 2015. www.facebook.com/megatrendsinbrutality (accessed April 22, 2015).

Johnson, Christopher. "Bricoleur and Bricolage: From Metaphor to Universal Concept." *Paragraph* (Edinburgh University Press) 35, no. 3 (2012): 355–372.

Kaplan, Louise J. *Cultures of Fetishism.* New York: Palgrave Macmillan, 2006.

Katz, Mark. *Capturing Sound : How Technology Has Changed Music.* Berkeley: University of California Press, 2004.

Kelley, Norman. *R&B, Rhythm and Business: The Political Economy of Black Music.* New York: Akashic Books, 2005.

Kenney, William Howland. *Recorded Music in American Life: The Phonograph and Popular Memory, 1890–1945.* ebook. New York, Oxford: Oxford University Press, 1999.

Knopper, Steve. *Appetite for Self-Destruction: The Spectacular Crash of the Record Industry in the Digital Age.* New York: Free Press, 2009.

Koepple, Kate. "Interview: Sarah O'Holla of My Husband's Stupid Record Collection." *Hyper-Organized.* Kate Koepple Design. October 13, 2014. http://recorddividers.tumblr.com/post/99929625572/interview-sarah-oholla-of-my-husbands-stupid (accessed June 27, 2015).

Kornelis, Chris. "Why CDs May Actually Sound Better than Vinyl." *LA Weekly.* January 27, 2015. http://www.laweekly.com/music/why-cds-may-actually-sound-better-than-vinyl-5352162 (accessed February 10, 2015).

Kozinn, Allan. "Have You Heard the New Caruso? (No Kidding)." The *New York Times.* February 20, 2000. http://www.nytimes.com/2000/02/20/arts/have-you-heard-the-new-caruso-no-kidding.html (accessed June 6, 2015).

———. "Long and Winding Road, Newly Repaved." The *New York Times.* September 2, 2009. http://www.nytimes.com/2009/09/06/arts/music/06alla.html?pagewanted=all&_r=0 (accessed June 6, 2015).

———. "Shades of Yesterday: New Vinyl Versions of Beatles' Albums on the Way." The *New York Times.* September 27, 2012. http://artsbeat.blogs.nytimes.com/2012/09/27/shades-of-yesterday-new-vinyl-versions-of-beatles-albums-on-the-way/ (accessed June 6, 2015).

———. "Weaned on CDs, They're Reaching for Vinyl." The *New York Times.* June 9, 2013. http://www.nytimes.com/2013/06/10/arts/music/vinyl-records-are-making-a-comeback.html?_r=0 (accessed May 2, 2014).

Kramer, Lawrence. *Why Classical Music Still Matters.* Berkeley, CA: University of California Press, 2007.

Kubernik, Harvey. "The Basement Tapes Complete." *Record Collector News.* December 11, 2014. http://recordcollectornews.com/2014/12/bob-dylan/ (accessed March 8, 2015).

Learn More About Mastered for iTunes. iTunes Store. Prod. Apple.

Lefebvre, Sam. "Four Men With Beards: Rescuing Classic Out-of-Print Releases from the Archives." *SF Weekly.* March 30, 2012. http://www.sfweekly.com/shookdown/2012/03/30/four-men-with-beards-rescuing-classic-out-of-print-releases-from-the-archives (accessed April 24, 2015).

———. "Record Peddler: Speculators Make a Mockery of Record Store Day." *SF Weekly.* April 23, 2013. http://www.sfweekly.com/shookdown/2013/04/23/record-peddler-speculators-make-a-mockery-of-record-store-day (accessed May 3, 2015).

Lendino, Jamie. "Apple Releases 'Mastered for iTunes,' but Sticks with Compressed Files." *PCMagazine.com.* February 23, 2012. http://www.pcmag.com/article2/0,2817,2400629,00.asp (accessed February 8, 2015).

Lerman, Nina E., Ruth Oldenziel, and Arwin P. Mohun. *Gender and Technology: A Reader.* Baltimore, MD: Johns Hopkins University Press, 2003.

Levine, George. *The Realistic Imagination: English Fiction from Frankenstein to Lady Chatterly.* Chicago: University of Chicago Press, 1983.

Levy, Joe. "Billboard Cover: Jack White on Not Being a 'Sound-Bite Artist,' Living in the Wrong Era and Why Vinyl Records Are 'Hypnotic.'" *Billboard.* March 6, 2015. http://www.billboard.com/articles/news/6494415/jack-white-billboard-cover-talks-third-man-records-white-stripes-elvis-presley (accessed April 23, 2015).

Lewry, Fraser. "Creedence Clearwater Revival: The Complete Studio Albums." *Classic Rock*, March 2015: 97.

Liberman, Mark. "Peeve Emergence: The Case of 'Vinyls.'" *Language Log.* June 12, 2012. http://languagelog.ldc.upenn.edu/nll/?p=4017 (accessed March 5, 2015).

Llewellyn Smith, Caspar. 2010. "Why the Beatles sealed the digital deal with iTunes." *The Guardian.* November 20. Accessed December 13, 2015. http://www.theguardian.com/music/2010/nov/21/beatles-itunes-apple-downlaods.

Lukács, Georg. "Reification and the Consciousness of the Proletariat." In *History and Class Consciousness*, translated by Rodney Livingstone, 83–222. Cambridge, MA: MIT Press, 1971.

Luke, Timothy W. *Museum Politics : Power Plays at the Exhibition.* Minneapolis: University of Minnesota Press, 2002.

MacDonald, Ian. "Turn on, Tune In . . ." *Mojo*, 2012. The Beatles Magical Mystery Tour Special Edition ed.: 8–15.

Maerz, Melissa. "The Great Folk-Rock Revival." *Entertainment Weekly*, March 1, 2013: 38–45.

Magnum, Theresa. "Dog Years, Human Fears." In *Representing Animals*, edited by Nigel Rothfels, 35–47. Bloomington: Indiana University Press, 2002.

Magoun, Alexander Boyden. "Shaping The Sound of Music: The Evolution of the Phonograph Record, 1877–1950." 2000.

Maines, Rachel. *Hedonizing Technologies: Paths to Pleasure in Hobbies and Leisure.* Baltimore: Johns Hopkins University Press, 2009.

Male, Andrew. "5 Beatles Songs That Sound Better In Mono." *Mojo* magazine July 8, 2014. http://www.mojo4music.com/15569/5-beatles-songs-that-sound-better-in-mono/ (accessed June 13, 2015).

Marsh, Dave. *The Beatles' Second Album.* New York: Rodale, 2007.

Martin, Andrew. "The Album Re-Release and Deluxe Edition Phenomenon." *Complex.* November 12, 2012. http://www.complex.com/music/2012/11/the-album-re-release-and-deluxe-edition-phenomenon (accessed April 15, 2015).

Marx, Karl. *Capital, Volume I.* London: Penguin Classics, 1990.

———. *Capital: A Critique of Political Economy, vol 1.* New York: Vintage, 1977.

"McIntosh." *Hi Fi Lit.* Ken Hagelthorn. (accessed January 28, 2015).

Mendelsohn, Jason, and Eric Klinger. "Counterbalance: The Beach Boys' 'SMiLE.'" *PopMatters.* July 25, 2014. http://www.popmatters.com/post/183988-counterbalance-the-beach-boys-smile/ (accessed March 11, 2015).

Milano, Brett. *Vinyl Junkies: Adventures in Record Collecting.* New York: St. Martins, 2003, 2012.

Millard, Andre. *America on Record: A History of Recorded Sound.* Cambridge and New York: Cambridge University Press, 2005.

Miller, Leonardo, and Chris Perez. "vibing." *Urban Dictionary.* June 12, 2008. http://www.urbandictionary.com/define.php?term=vibing (accessed June 30, 2015).

Miller, Scott. *Sex, Drugs, Rock & Roll, and Musicals.* Lebanon, NH: Northeastern University Press, 2011.

Mills, Sara. *Discourse.* London and New York: Routledge, 1997, 2004.

Moist, Kevin M. "Record Collecting as Cultural Anthropology." In *Contemporary Collecting: Objects, Practices, and the Fate of Things*, edited by Kevin M. Moist and David Banash. Lanham, MD: Scarecrow, 2013.

Moren, Dan. "The Beatles arrive on iTunes." *Macworld.* November 6, 2010. http://www.macworld.com/article/1155735/beatles_itunes.html (accessed June 8, 2015).

Morley, David. "What's 'Home' Got to Do with It: Contradictory Dynamics in the Domestication of Technology and the Dislocation of Domesticity." Edited by Thomas Berker, Maren Hartmann, Yves Punie and Katie Ward. Maidenhead, Berkshire: Open University Press, 2006.

Music Advertisements of the 1910s. 2010. http://www.vintageadbrowser.com/music-ads-1910s (accessed January 28, 2015).

Napier-Bell, Simon. *Ta-Ra-Ra-Boom-De-Ay: The Business of Popular Music.* Kindle. London: Unbound, 2014.

Nat, Happy. "Collectors Corner—Yesterday and Today (and the infamous "Butcher Cover")." *The Beatles Rarity*. January 30, 2013. http://www.thebeatlesrarity.com/2011/01/30/collectors-corner-yesterday-and-today-and-the-infamous-butcher-cover/ (accessed March 18, 2015).

Neal, Meghan. 2015. "The Little-Known Recording Trick That Makes Singers Sound Perfect." *Motherboard*. Vice Media LLC. December 1. Accessed December 10, 2015. http://motherboard.vice.com/read/the-little-known-recording-trick-that-makes-singers-sound-perfect.

"Notes on Harry Smith's Anthology." *Anthology of American Folk Music*. Smithsonian Folkways Recordings, 1997.

Oatman-Stanford, Hunter. "Meet the Irreverent Librarian Who's Taking on the Music Nerds." *Collectors Weekly*. April 24, 2014. http://www.collectorsweekly.com/articles/my-husbands-stupid-record-collection/ (accessed June 29, 2015).

O'Brien, Pat. "Letters to the Editor." *Billboard*, May 26, 1979.

O'Holla, Sarah. "A Certain Ratio 'Early.'" *My Husband's Stupid Record Collection*. February 24, 2014. http://alltherecords.tumblr.com/post/77762296856/a-certain-ratio-early. (accessed June 24, 2015).

———. "Adam and the Ants 'Prince Charming.'" *My Husband's Stupid Record Collection*. February 28, 2014. http://alltherecords.tumblr.com/post/78174521582/adam-and-the-ants-prince-charming.

—. "Song Hits of 1939." *My Husband's Stupid Record Collection*. February 22, 2014. http://alltherecords.tumblr.com/post/77528015983/song-hits-of-1939 (accessed June 27, 2015).

———. "The Chi Lites S/T (1973)." *My Husband's Stupid Record Collection*. September 11, 2014. http://alltherecords.tumblr.com/post/97267324454/the-chi-lites-s-t-1973 (accessed June 28, 2015).

———. "What is this?" *My Husband's Stupid Record Collection*. February 22, 2014. http://alltherecords.tumblr.com/post/77511421124/what-is-this (accessed June 22, 2015).

Opie, John. *Virtual America: Sleepwalking Through Paradise*. Lincoln and London: University of Nebraska Press, 2008.

Oremus, Will. "The Hot New Audio Technology of 2014 Is ... Vinyl?" *Slate.com*. January 6, 2014.http://www.slate.com/blogs/future_tense/2014/01/06/vinyl_lp_sales_hit_22_year_record_in_2013_digital_music_sales_down_chart.html (accessed May 2, 2014).

Orville, Miles. 2014. *The Real Thing: Imitation and Authenticity in American Culture, 1880-1940*. Twenty-Fifth Anniversary edition. Chapel Hill: University of North Carolina Press.

Osborne, Richard. *Vinyl: A History of the Analog Record*. Farnham, Surrey: Ashgate Publishing, 2012.

Patke, Rajeev S. "Benjamin on Art and Reproducibility: The Case of Music." In *Walter Benjamin and Art*, edited by Andrew Benjamin, 185–208. London and New York: Continuum, 2005.

Peim, Nick. "Walter Benjamin in the Age of Digital Reproduction: Aura in Education: A Rereading of 'The Work of Art in the Age of Mechanical Reproduction.'" *Journal of Philosophy of Education* 41, no. 3 (2007): 363–380.

Perpetua, Matthew. "Digital Sales Eclipse Physical Sales for First Time." January 6, 2012. http://www.rollingstone.com/music/news/digital-sales-eclipse-physical-sales-forfirst-time-20120106 (accessed March 25, 2013).

Phillips, Marc. "The Vinyl Anachronist." *Perfect Sound Forever*. February 1998. http://www.furious.com/perfect/vinyl.html (accessed June 17, 2015).

"Please Please Me, Parlophone, PMC 1202." *The Beatles Collection*. July 7, 2011. http://thebeatles-collection.com/wordpress/2011/07/16/please-please-me-parlophone-pmc-1202-3/ (accessed June 9, 2015).

Prial, Dunstan. *The Producer: John Hammond and the Soul of American Music*. London: Macmillan, 2007.

punktexas. "The Vinyl Community Record Shelves Yearbook 2013." *YouTube*. April 12, 2013. https://www.youtube.com/watch?v=0aONGI8QupY (accessed June 29, 2015).

R, Chris. "Record Collector U.K./Goldmine U.S.—Vinyl grading systems." *Steve Hoffman Music Forums.* September 12, 2003. http://forums.stevehoffman.tv/threads/record-collector-u-k-goldmine-u-s-vinyl-grading-systems.21433/#post-11026401 (accessed March 22, 2015).

Raile, Dan. "'High-Definition' Music Explained: Can You Really Tell the Difference?" *Billboard* January 7, 2015. http://www.billboard.com/articles/business/6429580/what-is-high-definition-music-debate-sony-walkman-pono (accessed February 4, 2015).

Rawson, Eric. "Perfect Listening: Audiophilia, Ambiguity, and the Reduction of the Arbitrary." *The Journal of American Culture* 29, no. 2 (June 2009): 202–212.

"Record Grading 101: Understanding The Goldmine Grading Guide." *Goldmine.* June 22, 2010. http://www.goldminemag.com/collector-resources/record-grading-101 (accessed March 22, 2015).

Record Store Day. "About Us." 2015. http://www.recordstoreday.com/CustomPage/614 (accessed April 30, 2015).

"Records My Cat Destroyed." *Tumblr.* June 28, 2015. http://recordsmycatdestroyed.tumblr.com/ (accessed June 29, 2015).

Reynolds, Simon. *Retromania: Pop Culture's Addiction to Its Own Past.* New York: Faber and Faber, 2011.

RHCD. "4 Men With Beards Label has Bad Pressings." *Steve Hoffman Music Forums.* February 15, 2014. http://forums.stevehoffman.tv/threads/4-men-with-beards-label-has-bad-pressings.343547/page-2 (accessed April 24, 2015).

Richardson, Mark. "Does Vinyl Really Sound Better?" *Pitchfork.* July 29, 2013. http://pitchfork.com/thepitch/29-vinyl-records-and-digital-audio/ (accessed February 10, 2015).

———. "The Enduring Cultural Weight of the Anthology of American Folk Music." *Pitchfork.* April 17, 2014. http://pitchfork.com/thepitch/313-the-anthology-of-american-folk-music-as-physical-artifact/ (accessed March 12, 2015).

Riggs, Michael. "Must We Test? Yes, We Must!" *High Fidelity*, January 1989: 5.

Riley, Tim. *Tell Me Why: A Beatles Commentary.* Boston: Da Capo Press, 2002.

Rivadavia, Eduardo. "The History of the New Wave of British Heavy Metal." *Ultimate Classic Rock.* May 8, 2014. http://ultimateclassicrock.com/new-wave-of-british-heavy-metal/ (accessed February 23, 2015).

Rockwell, Ken. "What is an Audiophile?" *Ken Rockwell.* October 2012. http://www.kenrockwell.com/audio/audiophile.htm (accessed June 3, 2015).

Rohter, Larry. "A Menagerie of Music Lives in a Box: Jack White Explores History of Paramount Records." *The New York Times.* October 25, 2013. http://www.nytimes.com/2013/10/27/arts/music/jack-white-explores-history-of-paramount-records.html?pagewanted=all (accessed March 27, 2015).

Rosen, Jay. "The People Formerly Known as the Audience." In *The Social Media Reader*, edited by Michael Mandiberg, 13–16. New York: New York University Press, 2012.

Rosen, Philip. *Change Mumified: Cinema, Historicity, Theory.* Minneapolis: University of Minnesota Press, 2001.

Rosoff, Matt. "Pet Peeves with the Vinyl Resurgence." *CNET.* June 17, 2010. http://www.cnet.com/news/pet-peeves-with-the-vinyl-resurgence/ (accessed March 25, 2013).

Schroeder, Patricia R. "Passing for Black: Coon Songs and the Performance of Race." 33, no. 2 (June 2010): 139–153.

Sørensen, Knut H. "Domestication: The Enactment of Technology." In *Domestication of Media and Technology*, edited by Thomas Berker, Maren Hartmann, Yves Punie, and Katie Ward. New York, NY: Open University Press, 2006.

Shankland, Stephen. "Sound Bite: Despite Pono's Promise, Experts Pan HD Audio." *CNET.com.* March 20, 2014. http://www.cnet.com/news/sound-bite-despite-ponos-promise-experts-pan-hd-audio/ (accessed February 10, 2015).

Sherry, John F. Jr., "Advertising As a Cultural System." In *Marketing and Semiotics: New Directions In the Study of Signs for Sale*, edited by Donna Jean Umiker-Sebeok, 451–460. Berlin: Walter de Gruyter & Co., 1987.

Shirley, Ian. "Help Me Rhonda." *Record Collector*, January 2015: 11.

Shuker, Roy. *Wax Trash and Vinyl Treasures: Record Collecting as a Social Practice.* Farnham, Surrey: Ashgate, 2010.

Silke, John. *Record Collecting in the Digital Age.* Amazon Digital Services, 2013.

Silverstone, Roger. "Domesticating Domestication: Reflections on the Life of a Concept." In *Domestication of Media and Technology*, by Thomas Berker, Maren Hartmann, Yves Punie, and Katie Ward. Maidenhead, Berkshire: Open University Press, 2006.

Sisario, Ben. "EMI Is Sold for $4.1 Billion in Combined Deals, Consolidating the Music Industry." The *New York Times.* November 11, 2011. http://www.nytimes.com/2011/11/12/business/media/emi-is-sold-for-4-1-billion-consolidating-the-music-industry.html (accessed June 6, 2015).

Sisario, Ben, and Miguel Helft. "Working It Out, iTunes to Sell Beatles Titles." *The New York Times.* November 15, 2010. http://www.nytimes.com/2010/11/16/business/media/16apple.html (accessed June 6, 2015).

Skiner, S. "Reader's Forum." *High Fidelity*, May 1955.

Slatoff-Burke, Megan. "The Legacy of the Camera Obscura." *Discoveries* (John S. Knight Institute for Writing in the Disciplines) Spring, no. 6 (2005): 29–34.

Sliwicki, Susan. "Go Beyond Book Values when Buying, Selling and Valuing Records." *Goldmine.* November 7, 2013. http://www.goldminemag.com/article/record-prices-and-values-are-not-the-same-thing (accessed March 24, 2015).

"SME—SME Model 30/12—includes Series V-12 tonearm." *Acoustic Sounds.* http://store.acousticsounds.com/d/94257/SME-SME_Model_3012_-_includes_Series_V-12_tonearm-Turntables (accessed June 3, 2015).

Souvegnier, Todd. *The World of DJs and Turntable Culture.* Milwaukee, WI: Hal Leonard Corporation, 2003.

Spizer, Bruce. *The Beatles' Story on Capitol Records, Part One: Beatlemania and the Singles.* New Orleans, LA: 498 Productions, 2000.

———. *The Beatles' Story on Capitol Records, Part Two: The Albums.* New Orleans, LA: 498 Productions, 2000.

———. "The Beatles: The Capitol Albums, Vol. 2." Capitol Records, 2006.

Stein, Erik. "Erik Stein: The Synth-Punk Sleuth." *Record Collector*, January 2014: 12.

Steffen, David J. 2005. *From Edison to Marconi: The First Thirty Years of Recorded Music.* Jefferson, NC and London: McFarland & Company, Inc.

Sterling, Scott T. "Tupac Shakur's Mom Sues Death Row for $1.1 Million and Unreleased Recordings." *Radio.com.* September 26, 2013. http://radio.com/2013/09/26/tupac-shakurs-mom-sues-death-row-for-1-1-million-and-unreleased-recordings/ (accessed April 15, 2015).

Sterne, Jonathan. *MP3: The Meaning of a Format.* Durham, NC: Duke University Press, 2012.

———. *The Audible Past: Cultural Orgins of Sound Reproduction.* Durham and London: Duke University Press, 2003.

Stoddard, Katy. "Ghosts of Christmas Past: Festive Adverts in the Guardian—in Pictures." *The Guardian.* December 21, 2011. http://www.theguardian.com/theguardian/from-the-archive-blog/gallery/2011/dec/21/christmas-past-adverts-guardian-archive#/?picture=383640150&index=5 (accessed January 28, 2015).

Stone, Steven. "The History of High-End Audio." *Audiophile Review.* Luxury Publishing Group Inc. September 8, 2011. http://audiophilereview.com/the-history-of-high-end-audio.html (accessed June 4, 2015).

Stooksbury, Clark. "Meet the Vinyl Community." *Reason.com.* November 2013. http://reason.com/archives/2013/10/19/meet-the-vinyl-community (accessed June 29, 2015).

Stormo, Roger. "The Beatles 'Movie Medley' 2010 Upgrade." *WogBlog—All Things Beatle!* October 8, 2010. http://wogew.blogspot.com/2010/10/beatles-movie-medley-2010-upgrade.html (accessed June 7, 2015).

Strangelove, Michael. 2010. *Watching YouTube: Extraordinary Videos by Ordinary People.* Toronto, Buffalo, London: University of Toronto Press.

Straw, Will. "Music as Commodity and Material Culture." *Repercussions* 7–8, Spring–Fall (1999–2000): 147–171.

Stubbs, David. "The Beatles Mono Vinyl Box Set." *Classic Rock*, October 2014: 97.

Sturken, Marita, and Lisa Cartwright. *Practices of Looking: An Introduction to Visual Culture.* Oxford University Press, 2009.
Suisman, David. *Selling Sounds: The Commercial Revolution in American Music.* Cambridge, MA: Harvard University Press, 2009.
Sullivan, Matt. "Reissue of the Year: Lewis L'Amore." *Mojo,* January 2015: 66.
Sweney, Mark. "Digital Music Spending Greater than Sales of CDs and Records for First Time." *The Guardian.* May 31, 2012. http://www.theguardian.com/media/2012/may/31/digital-music-spending-bpi (accessed March 25, 2013).
The Anstendig Institute, "Stereo: A Misunderstanding." *The Anstendig Insitute.* 1982, 1984. http://www.anstendig.org/Stereo.html (accessed April 29, 2015).
The Berliner Gramophone: A Gallery of Rare Advertisements (1896–1901), 2011. http://www.mainspringpress.com/berliner-ads.html (accessed January 28, 2015).
"The Human Voice Is Human on the New Orthophonic Victrola," Camden, NJ: The Victor Talking Machine Company, 1927.
"The Law." *RIAA.* 2015. http://www.riaa.com/physicalpiracy.php?content_selector=piracy_online_the_law (accessed March 3, 2015).
The Poor Audiophile, "Bose and the Audiophile: The Joy of Disdain." *The Poor Audiophile.* November 8, 2013. http://www.pooraudiophile.com/2013/11/bose-and-audiophile-joy-of-disdain.html (accessed June 3, 2015).
"The Story Behind the Beatles U.S. Albums—Including Differrences Between U.S. and U.K. Mixes," *Guitar Afficianado.* December 26, 2013. http://www.guitaraficionado.com/the-story-behind-the-beatles-us-albums-including-differences-between-us-and-uk-mixes.html (accessed June 7, 2015).
Third Man Records. "Jack's Billboard Cover Story and Recent Acquisition." *Thrid Man Records.* March 6, 2015. http://thirdmanrecords.com/news/jacks-billboard-cover-story/ (accessed April 21, 2015).
Thomas, Chris. "Afeni Shakur Promises the Release of Tupac's 'Entire Body of Work'" *HIPHOPWIRED.* February, 2013. http://hiphopwired.com/2013/02/18/afeni-shakur-promises-the-release-of-tupacs-entire-body-of-work/ (accessed April 15, 2015).
Thompson, Dave. "Dear Labels: Please Don't Skimp on New Vinyl Releases, Reissues." *Goldmine.* April 25, 2013. http://www.goldminemag.com/article/dear-labels-please-dont-skimp-on-new-vinyl-releases-reissues (accessed April 15, 2015).
Thompson, Erik. 2015. "Why Vinyl Is Resurgent in a Digital Age." *The Current.* November 27. Accessed November 29, 2015. http://blog.thecurrent.org/2015/11/why-vinyl-is-resurgent-in-a-digital-age/?WT.mc_id=41d68f68444d3ad3f3a307dfee9a5d32.
Trelease, Greg. "In Digital Age Vinyl Records Endure." *Masslive.com.* March 14, 2013. http://www.masslive.com/entertainment/index.ssf/2013/03/in_digital_age_vinyl_records_e.html (accessed March 25, 2013).
Turkle, Sherry. *Together Alone: Why We Expect More from Technology and Less from Each Other.* New York: Basic Books, 2012.
U.S. Pressing Plants. http://www.anorakscorner.com/Key.html (accessed March 20, 2015).
"Vaccine." *Lostpedia: The Lost Encyclopedia.* June 20, 2014. http://lostpedia.wikia.com/wiki/Vaccine (accessed May 10, 2015).
Vaher, Berk. "Identity Politics Reco(r)ded: Vinyl Hunrers as Exotes in Time." *TRAMES: A Journal of the Humanities and Social Sciences* 12, no. 3 (2008): 342–354.
Van Wyck Farkas, Remy. "Reader's Forum." *High Fidelity,* March 1954.
Vechinski, Michael James. "Collecting, Curating, and the Magic Circle of Ownership In a Postmaterial Culture." In *Contemporary Collecting: Objects, Practices, and the Fate of Things,* edited by Kevin M. Moist and David Banash. Lanham, Toronto, and Plymouth, UK: The Scarecrow Press, Inc., 2013.
Victor Talking Machine Co. "Instructions for the Setting Up, Operation and Care of the Victrola Spring Motor Type." Camden, NJ: Victor Talking Machine Company, June 27, 1924.
———. "Instructions for the Unpacking, Assembling, Operattion and Care of the Victrola XI." Camden, NJ: Victor Talking Machine Company, 1920.
Villchur, Edgar. "How to Get the Most from Your Loudspeakers." *Hi Fi/Stereo Review,* October 1961: 59–62.

Vonnegut, Kurt. *Player Piano.* New York: Delacorte, 1952.
Waehner, Michael. *Get In the Groove: A Begginer's Guide to Vinyl in the 21st Century.* Kindle Edition.
WALL-E. Directed by Andrew Stanton. Produced by Walt Disney Pictures; Pixar Animation Studios. Walt Disney Studios Motion Pictures, 2008.
Wayne, Michael. "The Most Comprehensive Record Cleaning Article Ever!" *Analogue Planet.* January 1, 2012. http://www.analogplanet.com/content/most-comprehensive-record-cleaning-article-ever-0 (accessed May 31, 2015).
Welcome to the Rare Record Price Guide Online! Metropolis International Group Ltd. 2010–2015. http://www.rarerecordpriceguide.com/ (accessed March 20, 2015).
"Why Do I Collect Records (Instead of CDs)?" *Analog Jazz.* September 15, 2013. Accessed December 1, 2015. https://analogjazz.wordpress.com/2013/09/15/why-do-i-collect-records/.
Williams, Alex. "On the Tip of Creative Tongues." The *New York Times.* The New York Times Company. October 2, 2009. http://www.nytimes.com/2009/10/04/fashion/04curate.html?pagewanted=all&_r=0 (accessed April 24, 2015).
Williams, Raymond. *Marxism and Literature.* Oxford: Oxford University Press, 1977.
Willis, Susan. "Memory and Mass Culture." In *History and Memory in African-American Culture*, edited by Geneviéve Fabre and Robert O'Meally, 178–187. New York, Oxford: Oxford University Press, 1994.
Wilson, David Bertrand. "The Beatles." *Wilson and Alroy's Record Reviews.* http://www.warr.org/beatles.html (accessed June 13, 2015).
Windmüller, Sonja. "Trash Museums: Exhibiting in Between." In *Trash Culture: Objects and Obsolescence in Cultural Perspective*, edited by Gillian Pye, 39–58. Bern, 2010.
Wolff-Mann, Ethan. 2015. "Vinyl Record Revenues Have Surpassed Free Streaming Services Like Spotify." *Time.com.* October 1. Accessed January 2, 2016. http://time.com/money/4056464/vinyl-records-sales-streaming-revenues/.
Womack, Kenneth. "Authorship and the Beatles." *College Literature* 34, no. 3 (Summer 2007): 161–182.
———. *The Beatles Encylcopedia: Everything Fab Four, Volume 2: K–Z.* Santa Barbara, CA: Greenwood, 2014.
Yochim, Emily Chivers, and Megan Biddinger. "'It kind of gives you that vintage feel:' Vinyl Records and the Trope of Death." *Media, Culture and Society* (SAGE Publications) 30, no. 2 (2008): 183–195.
YouTube Vinyl Community. Accessed December 31, 2015. https://www.facebook.com/groups/116959488388249/?fref=ts.
Žižek, Slavoj. "Cyberspace, Or the Virtuality of the Real." *Journal of the Centre for Freudian Analysis and Research.* http://www.jcfar.org/past_papers/Cyberspace%20and%20the%20Virtuality%20of%20the%20Real%20-%20Slavoj%20Zizek.pdf (accessed June 25, 2015).
Zolten, Jerry. "The Beatles As Recording Artists." In *The Cambridge Companion to the Beatles*, edited by Kenneth Womack, 33–62. Cambridge and New York: Cambridge University Press, 2009.
Zwicky, Arnold. "Countification." *Language Log.* August 8, 2008. http://languagelog.ldc.upenn.edu/nll/?p=501 (accessed March 5, 2015).

Index

Abbey Road, 28, 31, 43n4
abrasive, 123
absence, 32
accompaniment, 29, 109
accuracy, 120
acetate, 95, 96
Acoustic Sounds, 129
acoustical: acoustical environment, 120; acoustical recording, 6
acquisition, 70, 138, 149, 150
"The Active Audience", 140
activity, 62, 94, 97, 98, 107n1, 114, 116, 138
actor-network theory, 117
actualization, 36
actualized, 31, 41
ad, 9
Adam, 31; Adamic Myth, 31
adaptation, 53, 72
addiction, 152
administrators, 153
adoption, 114, 135
Adorno, Theodor, 24, 36, 39, 70, 72, 97
Adornoesque, 72
advertising, 1, 3, 4, 9, 16, 19, 74, 160; advertisements, 9, 51, 99
aesthetic: aesthetic claims, xv, 46; aesthetic discourse, xiv; Hi-fi aesthetics, 15; Lo-fi aesthetics, 15; No-fi aesthetics, 15; reproducibility of aesthetic objects, the, 24

affluence: affluent, 120, 128
aficionados, 41, 53, 102, 111, 114, 133
agency, 20, 56, 71
album, xi, xii, xiii, 13, 24, 25, 26, 27, 28, 30, 31, 33, 34, 37, 38, 39, 40, 41, 42, 45, 49, 52, 53, 55, 57, 59n2, 69, 76, 77, 78, 79, 81, 87, 88, 92, 93, 97, 98, 102, 124, 133, 146, 147, 155
alienated, 36, 67, 89
alienation, 89
All Music, 30, 77, 80
allegory, 145
Altschuler, Dick, 93
Amazon.com, 43n3, 45, 52, 92, 140
The American Hand Gramophone reproducer, 118
ambience, 30, 101
ambiguity, 4, 20, 21n2
Amoeba Records, 136
amplifier, xii, 9, 53, 120, 123, 127, 128, 129, 147
Amputechture, 73
analog: analog playback technologies, xv; analog recording, xiv, xv, 13, 35, 36, 39, 40, 47, 48, 52, 88, 116; analog sound technologies, xiii, xv; analog turntable, 49
Analog Jazz, 52
Analog Planet, 13, 125
Anderson, Jim, 18, 20
Anderson, Tim J., 120

"Angelus Novus", 65
Anorak's Corner, 82
The Anstendig Institute, 102
Antarctica, 102
Anthology of American Folk Music , 77
antitraditionalist, 40
anxiety, 6, 84, 85, 89, 94, 144
appearance, 9, 39, 48, 67, 89, 105
Apple, 25, 37, 39, 78
Apple Corps, 34, 37
Apple Records, 28
Apple Inc., 25, 26, 27, 34
application, 11, 125
appreciation, 40, 137, 154
archaic, 46, 57, 109, 149
archeology: archeological, 39, 89, 93; archeologists, 79
architecture, 129
archive: archival, 26, 74, 156; archiving, 71; Archontic, 156
arena, 63, 71
Ariston, 137
arrangement, 117, 120
artefacts, 16, 100, 134
artifact, 21n3, 32, 33, 37, 47, 62, 63, 87, 100, 135, 156
artifice, 36
artificiality, 19
artist: artistic, 40, 52, 63, 74, 77, 90, 96; artistry, 33; artwork, 26, 32, 39, 50, 98, 115
ascetic, 70, 111
assimilation, 7, 99; assimilated, 35, 115
Atkinson, John, 101
attachment, 106, 146, 156
Attali, Jacques, 34, 35, 43
auction: auctioned, 95
audience, 9, 16, 34, 37, 48, 49, 53, 59n2, 90, 120, 140
"audile technology", 101
audio engineer, 9, 14
audio output, 9
audiophilia, xv, 19, 21n2, 112, 122, 126, 127, 128, 129, 130n5, 151
auditor, 13
auditory, 101
aural, 46, 58, 129; auralessness, 35
auratic, 36, 39
authenticate, 147

authenticity, xiv, 8, 23, 32, 35, 36, 47, 48, 49, 50, 51, 52, 53, 55, 56, 57, 58, 96, 105, 143, 145
authorship, 32
automation, 117
autonomy, 36, 39, 67, 72, 74, 80, 98, 103, 150

Backhouse, Jim, 64
Bakardjieva, Maria, 153
Bakhtin, 72
Banash, David, 67, 139
The Band, 74
The Bangles, 106
bankruptcy, 90, 92
barbarism, 65
Barker and Yuval Taylor, Hugh, 48
Barnes and Noble, 153
baroque, 95
Bartmanski, Dominick and Ian Woodward, xiii, 49, 63, 90, 94, 95, 96, 97, 100, 102, 104, 105, 106, 115, 124, 142
bass, 9, 19, 41, 122, 133, 135, 159
bastardization, 29, 37
battery, 114
Baudrillard, Jean, 84, 85, 95, 96, 102
Baym, Nancy K., 142, 154, 155
Bazin, André, 7, 18, 114
BBC, 140
The Beach Boys, 76
The Beatles: *Anthology*: The Beatles' Cannon, 25; *Beatles In Mono*, 28; "Beatles Vinyl Update—Fantastic Grail.", 23, 43n1; "butcher cover", 79; "The Beatles' Movie Medley", 30; Rubber Soul, 26, 27, 37, 40, 79; *Sgt Pepper's Lonely Hearts Club Band*, 26, 28, 75; *VI*, 37
Beatlemaniac, 28
Beaulieu, 107n2
Beedle, Ashley, 63
BEETHOVEN, 13
Beidelman, T.O., 100
Belam, Martin, 25
Benjamin, Walter, xiv, 23, 24, 32, 34, 35, 36, 39, 40, 51, 55, 62, 65, 71, 72, 113
Bennett, Tony, 110
Berger, Thomas, 4

Berker, Thomas, et. al., *Domestication of Media and Technology*, 3
Berkowitz, Steve, 41
Berliner, Emile, 3, 4, 6
Berliner Gramophone, 8, 118, 119
Berraud, Francis, 1, 2, 7
Best Buy, 85
The Bicycle Thief, 18
Bikini Kill, 151
Bill Monroe, 77
Billboard, 13, 14, 28, 76, 96, 102, 137
binary, 145
biographies, 88
Black Swan Records, 90
blackface, 8, 66
Blender, Morton, 12, 13
blogs/bloggers, 26, 30, 52, 73, 79, 104, 126, 133, 136, 137, 140, 142, 143, 146, 147, 148, 156
Blu-ray disc, 85
bluegrass, 77
Blues, 33, 48, 65, 66, 77, 82, 91, 92
bodily, 6, 110, 112, 119
Bogen Sound Systems, 10, 11
Boggs, Dock, 77
Bon Iver, 57
bongo, 121
booth, 40, 59
bootleg, 63, 74, 75, 76, 77; bootlegged, 76; bootleggers, 74, 76; bootlegging, 74
Borgmann, Albert, 53, 54, 145
Bose, 128
Boston Audio Society, 18
boundary, 49
bourgeois, 34, 98, 101, 127
bourgeois art, 97
bourgeoisie, 34
boutique, 145
boyd, dana, 141
brain, 107n1, 130, 144
Bratich, Packer, and McCarthy, 110, 112
Brian Wilson Presents SMiLE, 76
Bricolage: Bricoleur, 80, 81, 156
Brittan Jr., Gordon G., 54
brittleness, 123
Broadway, 66
brochure, 9
Brogmann, 53
Bruck, Connie, 92

BSEE, 127
Buckeye Music Company, 8
bureaucracy, 97
Bürger, Peter, 97, 98
Burnett, Bucks, 99
Bush, John, 76, 77
Buy 'n' Large, 85

Cabinet of Curiosities, 95
Cabinet of Wonder, 88, 92
Cage, Nicolas, 28
Calamar and Gallo, 97, 106
camera obscura, 10, 11
campaigns, advertising, 9, 13, 67, 80
Campbell, Neil, 63
canine, 1, 2, 3
canon, 97
canonical, 25, 97, 118
Canterbury scene, 80
capital, bridging, 155
capitalism, 33, 67, 72, 89, 102, 103, 104, 105; capitalist, 68, 69, 70, 72, 103, 105, 156; industrial capitalism, 25, 58, 61, 67, 70, 71, 72, 80, 81, 85, 89, 95, 97, 99, 118
Capitol Records, 30, 37, 41, 76, 78, 79
Caro, Mark, 41
cartridge, 129
Caruso, Enrico, 6, 28, 29, 31, 43n3
cassette, 25, 135
Cavell, Stanley, 62, 65
Cavern performance, 40
CD, xi, xii, 13, 14, 15, 16, 19, 25, 26, 27, 28, 36, 37, 41, 46, 47, 48, 53, 55, 69, 71, 76, 77, 78, 98, 115, 123, 135, 137
Celestial Jukebox, 160
Cerra, Allison, 143
The Chi Lites, 146
The *Chicago Tribune*, 41
Christgau, Robert, 77
chronotope, 36
cinema, 7, 18, 51, 114
circuit, 158
civilization, 65
Classic Rock magazine, 87
Cloud, the, 46, 56
Cobain, Kurt, 48
Coercive, 119, 122, 128
Coffin, 63, 88

cognitive mapping, 145
Coliseum, Oakland, 87
collages, 40
collecting: fetichistic, 63; systematic, 63
Collection, record, x, xiv, xv, 23, 41, 42, 43n2; Collectable, 42; collecting practices, xiv; vinyl record collectors, xiii, xiv, xv, 27, 33, 42
Collins, Arthur, 8
Colliton, Carrie, xiii
colonization, 71, 92
Coltrane, John, 97
Columbia House, 136
Columbia Records, 9, 93
Commodity: "commoditization of experience", 34; commodity fetishism, 33, 88, 92, 95, 96, 97, 99; fetishized commodities, xiv, 100, 106
competitive individualism, 140, 153
comping, 55
complaint, Morton Blender's, 12, 13
component, audio system, 11, 119, 127, 129
composer, 10, 13, 21n2, 91
composite "performance", 55
composition, 21n2
compression, 55
computer, 109, 110, 128, 144, 145
concert, 10, 15, 33, 53, 59n2, 63, 93, 120, 121
conformity, 64, 122, 147
Connery, Sean, 28
connoisseurship, 63, 112
Constantinides, John, 82
containers, records as, 37, 95, 100, 106
contextualization, 67, 80, 142
controversy, 14, 27, 37, 72, 84
Cook, Deborah, 71, 72
coon songs, 8, 66
corporal, 122
counter-discourse, 61, 76, 85; counter-discursive, 77, 156
counter-histories, 67
countification, 73
crate-diggers, xiii, 135
crate-digging, 151, 155
Creedence Clearwater Revival, 87
Creme, Lol and Kevin Godley, 111

criticism, 1, 2, 6, 7, 21n2, 30, 31, 39, 41, 42, 62, 72, 100, 147, 150, 153, 157
crooners, 121
Crosley, 128, 153
Crumb, R., 87
cueing, 135
cultural studies, xiii
culture industry, 24, 72
curate: curating, 80, 107n4, 140; curation, 39, 41, 79, 80, 87, 97, 98, 100; curatorial, 98, 100, 139
Curate Couture, 98
Currie, Mark, 31
Currin, Grayson, 88
Cutrona and Russell, 155
cyber, 144, 145, 148, 156, 157; cyberculture, 139; cybernauts, 157; cybersea, 140; cyberspace, xv, 49, 133, 137, 138, 139, 144, 145, 147, 148, 149, 156
cylinder, 3, 5, 7, 12, 16, 66, 99, 149

D'Onofrio, Steven, 74
Daston and Galison, 11
data: database, 79; microdata, 138
Dave Matthews Band, 104
Davis, John, 47
Davis, Jonathan, 39, 40
Davis, Miles, 25, 151
de Certeau, Michel, 99
deadwax, 81, 82
dealer, 83, 122, 124
Dean, Johnny, 81
Dean, Mitchell, 109, 111
Death Row, 92
debased commodities, 70
Decca Records, 12
Decca Record Brush, 114
DeChaine, D., Robert, 139
decoding, 81
deconstructing, 7
defects, 12, 83
deficiencies, 113
degradation, 27, 52; degraded, 102; degrades, 27
dehumanizing, 50
deluxe reissue, 87, 93, 94, 100
democratic, 153
democratize, 156; democratization, 138

demonstration, 117, 122, 149
deontology, 111
dependence, 54
dependency, 89
depreciation, 32
Depression, The Great, 16
Derecho, Abigail, 156
Derrida, Jacques, 31, 156; *Of Grammatology*, 31
Desmond (*Lost*), 109, 110, 130
determinant, 18, 19, 83
devalue, 37
developers, 8
device, 53, 54, 106, 111, 112, 114, 119, 120, 124, 141
device paradigm, the, 53–54
devotion, 47
Dexter, Dave, 30, 37, 78; Dexterization, 30, 41
The Dharma Initiative, 109
dialectical, 18, 36, 146
dialectical tension, 62
dialog, 146
diaphragm, 117
Dickerson, Althea, 91
diffusion, 102, 117, 145
digital audio, 14, 55
digital audio workstation (DAW), 55
digital compact disc, 9
digital distribution, xiii, 26
digital downloads, xi, 27, 53, 70, 105
digital media, xv, 45, 46
digital stores, 45
digitalization, 102
digitally sourced, 13
digitizing, 27
direction (of sound), 101, 102, 120
discernment, 138
disciplinary, 112, 114, 118, 119, 122, 124, 126, 129
discographies, 77, 79, 80
Discogs, 80, 102, 136, 139, 140
discourse community, 13
discursive practices, xiii, xiv, 11, 19, 33, 65, 68, 72, 74, 75, 77, 79, 85, 112, 119, 137
discriminating, 121
Discwasher, 114, 127
disembodied, 6, 28, 45, 49

disenfranchised, 4
disjunctions, 16
Disney Corporation, 85
disorder, 62, 65
disposability, 67
disposable, 79, 105
disruption, 109
disruptive, 37
disseminated, xiv, 25, 28, 105, 122
disseminates, 133
disseminating, 77
dissident values, 51
distortion, 15, 18, 19, 113, 123
distribution, 34, 88, 89, 97
DIY, 57, 65, 68
documentary recording, 16
Domestication Theory, 2, 3. *See also* actor-network theory
Douglas, Alan, 38
Downbeat magazine, 93
download; see digital distribution; digital downloads; digital media; digital stores
DRM, 25
dual, 137
duality, 62
Dunn, Robert G., 88, 89
Duophonic stereo, 30
Dunn, Robert G., 88, 89
Durrans, Brian, 100
Dutch prog, 80
DVD, xi, 85, 137
Dylan, Bob, 74, 75, 76, 77, 151
Dylanologist, 75
dynamic range, 41, 55

earbuds, 54, 56
eBay, 23, 45, 102, 103, 136
ecommerce, 140
ecosystem, 128, 138
Ed Sullivan Show, 34
Edison, Thomas, 3, 4, 5, 7, 9, 16, 56, 99, 135, 149. *See also* tinfoil
Edison Realism Test, 5, 6, 112
educators, 73
eight-track, 99
The Eight Track Museum, 99
Eisenberg, Evan, xiii, 70, 71, 90, 113, 157, 159, 160

Elborough, Travis, xiii, 49, 53, 55
electrical recording process, 9
electricity, 9
electromagnetic, 109
electroplating, 81
Eliot, Cass, 109
emancipatory, 156
emergent, 56, 57
Emerick, Geoff, 30
EMI Records, 25, 26, 27, 28, 30, 37, 38, 42, 81
empowerment, 138; empowering, 138, 156
emptiness, 53
enculturation, 56, 109, 110
enhancement, 140
enlightenment, 34
entertainment, 6, 56, 89, 92, 133, 145
Entertainment Weekly, 57
entrepreneurs, 3, 4, 16, 149
environment, 120
episteme, 56; epistemological, 28, 100; epistemological claims, xv
EPs, 25
equalization, 41; of works of art, 40
equipment, 3, 9, 11, 12, 13, 15, 16, 18, 19, 47, 113, 114, 120, 122, 123, 124, 127, 128, 129, 130n5, 147, 150, 151, 160
equivalence, 114; equivalencies, 145
Erlewine, Stephen, 78
escapism, 145
Esquivel, Juan Garcia, 32
ethics, 141
ethnicities, 151
ethnography, 152; ethnographic, 65, 100, 135
Etsy, 57, 149
Evans, Gil, 151
Evans, Walker, 51
EVE, 61
eventalization, 112
excavation, 39, 41, 89
excess of energy, 104
"exclusion of the new", 24
exclusivity, 82, 103
exhibition, 19, 98, 100, 135
exigency, 84
exotic, 58, 65; exotes, 64; exoticism, 65
experimentation, 128

expertise, 104, 105, 112, 128, 129, 143, 147, 156
exploitation, 9, 66, 67, 68, 69, 70, 90, 91, 92, 93, 103, 104, 105, 119
exploration, xv, 143
expropriating, 72

fabricated, 100
Facebook, 79, 95, 106, 107n2, 136, 140, 141, 149, 150, 151, 153, 158n2
facsimile, 50, 93, 95, 101
factory, 42, 83, 89, 93, 116
Fahey, John, 77
faithfulness, or reproduction, 2, 10, 15, 16, 35, 39, 53, 55, 95, 98, 101, 137
Faking It: The Quest for Authenticity in Music, 48
fallibility, 115, 123
falsification, 89, 96, 102, 112
fandom, 57
fantasy, 58, 141, 145, 146
Fantel, Hans H., 123
fanzines, 63, 79, 81
Farberman, Brad, 99
Farkas, Remy Van Wyck, 12, 13
farm-to-table, 57
Farmer, Nick, 31, 42, 43n2, 55, 81
Feedbands, 57, 58, 136
Feenberg, Andrew, 153
Feldman, Karen, 39
Felix, 133, 158n1
feminizing, 128
Ferreday, Debra, 143, 145, 146
fetish,: fetishistic, 15, 63; fetishization, 95, 104, 106
FFRR or full frequency range recordings, 12
filter, 18, 19, 62, 113
first pressing, 23, 55, 64, 79, 81
Fitzpatrick, Rob, 31
FLAC, 19
flippers, 104, 106
flipping, 103
fluid, cleanng, 125
focal practices, 54
focal things, 53, 54
Fogerty, John, 87
folk, 77; folkies, 77; folkloric, 95; folklorists, 48, 49; the folk-rock revival,

57, 58
Folkways Records, 77
format, music, xii, 14, 16, 25, 26, 27, 30, 46, 47, 51, 52, 57, 62, 72, 81, 99, 109, 135, 137, 152, 159, 160
formulation, 17, 24, 31, 52
forums, online discussion, 47, 55, 73, 84, 98, 140, 153
Foucault, Michel: Foucauldian, xv, 5, 11, 109, 116, 117, 122; Foucault, 4, 11, 13, 71, 110, 112, 117, 118, 119, 122, 125, 126, 129
foundation, xiii, xv, 35, 66, 127
fragility, 3, 109, 114, 124
Frankenstein, 38
Frankfurt School, 72, 89
Fremer, Michael, 126, 131n7, 137. *See also* Analog Planet
frequency : frequency response, 9, 18, 120; recorded frequencies, 16
Freud, 68; Freudian, 145
fuzz, 114

Gadamer, Hans-Georg, 1
Garcia, Jerry, 77
Garland, Emma, 103
Garnett, Robert, 72
Garrard, 137
gatefold, 87, 133, 147
the gaze, 4
gear, 41, 127, 128, 137
gender, xv, 120, 127, 146, 150, 151; *Gender and Technology: A Reader*, 127; gender relations, 143; gender stereotype, 147; gender tropes, 143; gendered, 142
genre, 48, 64, 65, 68, 72, 80, 97, 98, 122, 138, 152, 154
genuine, 16, 103
Georgetown Records, 135
The GI Bill, 111
Gilbert, Pat, 28, 30
gimmick records, 121
Gitelman, Lisa, xiii, 7, 8, 35, 135, 137; *Scripts, Grooves, and Writing Machines*, xiii
Glasgow, Joshua, 15, 16, 17
Gold, Gary Pig, 75
Goldman, Alex, 142, 143, 146

Goldmine, 62, 83, 84, 85, 94
Goodreads, 139–140
Gould, Glenn, 55
government, 110–112
governor, 119
The Governor (*The Walking Dead*), 45
Grail, 23, 43n1
Grajeda, Tony, 120, 129
grammarians, 72
gramophone, 1, 2, 3, 6, 7, 9, 12, 28, 118–119. *See also* Victrola; Victrola XI Talking Machine
The Great Depression, 16
Greenblatt, Stephen, 142
grief, 2; grieving, 1
grievances, 153
groove, 7, 15, 29, 32, 45, 52, 55, 71, 81, 88, 92, 93, 96, 105, 106, 113, 116, 117, 118, 123, 124–125, 127, 131n7. *See also* microgroove long player
grunge, 48
guitar, 27, 107n3; guitarist, 87

Habermas, Jürgen, 71–72
habit, 81, 128, 130, 139, 146, 147; habitus, 112
Haley, John, 28, 41
Hall, Stuart, and Nikolas Rose, 112
ham radio, 128, 130n5
Hammond, John, 78, 93
Hand, Martin, 138
handicrafts, 149
hangout, 106
hardware, 116
Harley, Robert, 42–43
Harmon International, 16
harmonica, 121
harmony, 34
Harrison, George, 34
Hayden, 13
Hayes factory, 42
Hayes, David, 67, 70, 72, 79
HDtracks.com, 128
headphones, 147. *See also* earbuds
heavy-grade virgin vinyl, 94
hegemony, 66
Heidegger, Martin, 36, 53
Hello Dolly, 61
Hendrix, Jimi, 38

heritage, 88, 93–94; heritagization, 90
Hesmondhalgh, David, xiii, 140, 153, 154, 156
Hess, Amanda, 142, 143
heteroglossia, 72
Heylin, Clinton, 74
hi-res movement, 14
high fidelity, 9–13, 16–18, 29, 112, 119, 120, 123, 127, 128, 129, 130n5
High Fidelity , 9–10, 12, 122
hi-fi ideal, 15
Hi-Fi Stereo Review, 123
hi-fi system, 13
Hinson, Peter, 103
hip-hop, 135, 149
hipster, 46
historians, 4, 65, 127
hobby, 128–129; hobbyists, 122
Hoffman, Steve, 47, 84, 98
Hoffs, Susanna, 106
Hogan, Marc, 28
Holmes, Thom, 123, 124
Holt, J. Gordon, 15
home audio system, 129
Homer, Sean, 58
Hornby, Nick *High Fidelity*, 53
Hudson, Garth, 75
hum level, 9
human ear, 11, 21n3, 120, 122
human personality, 9
humanizing, 52. *See also* dehumanizing
hype, 6
hyperreality, 102

icon, 1, 8; iconic, 63; iconography, 4
idealization, 58
identification, 5, 82, 105, 128, 141
identity formation, 1, 2, 63, 110, 141, 148, 150
ideology, 8, 67, 127; ideological, 34, 65, 70, 100
Ikea, 133
illegal downloading, 36
imagination, 1, 18, 105, 106, 145
Imbach, Martin, 135
imitation, culture of, 50
immediacy, 52
immersion, 133
imperfect, 98, 114; imperfections, 42, 57, 84, 95, 114, 152
imperialism, 66
improvement of listeners, 3
improvement of technologies, 16
inauthentic, 27; inauthenticity, 75
incompleteness, 143
incomprehensibility, 156
inconvenient, 17, 66
indefinability of cyberspace, 149
independent record labels, 68, 95, 155
indestructible, 123, 124
indie, xiii, 46, 97, 102
individuality, 51
individuation, 101
industrialization, 34, 128
inequality, 66, 67
information age, the, 45, 49, 137, 156
information revolution, the, xv
information technology, xv
infrastructure, 138
infringement: patent, 7; copyright, 34
innovation, 3, 16, 25; innovators, 3
insecurity, 89
inspiration, 97
Instagram, 140
Institution, 117, 153
instructions, 5, 116–119, 123, 124, 125, 130n4
instrumentalized, 129
instruments (musical), 6
intangibility, 105
intellectual property, 37, 160
interactivity, 141
Internet, xv, 37, 45, 69, 79, 80, 133, 134, 135–136, 138, 139, 149, 150, 153, 156, 160
intertextual, 156
intimacy, 52, 101
invention, 8, 13, 35, 57, 136
inventors, 3
investment, 19, 73, 85, 123, 145, 154
iPhones, 134
iPod, 99, 134
"Mastered for iTunes", 17
iTunes Plus, 17

Jamaica, 82
James, Dick, 23

jazz, 48, 52, 66, 92, 97, 105, 138
Jobs, Steve, 34
John, Elton: *Goodbye Yellow Brick Road*, 55
Johnson, Clark, 59
Johnson, Eldridge, 3, 4
Johnson, Christopher, 80
Jones, Max, 124
journalists, 46
jukebox, 135, 149, 159
junk, 49–51

Kaplan, Louise J., 89
Katz, Mark, 5, 6, 17, 32, 71, 105, 127, 136, 160; *Capturing Sound: How Technology has Changed Music*, 17, 160
Kelley, Norman, 90, 91, 92
Kenney, William Howland, xiii, 3, 4, 6, 20, 90, 91
The Kingston Trio, 77
Klee, Paul, 65
Koepple, Kate, 143
Koetsu Blue Lace Platinum Magnet, 129
Kolster radios, 123
Kornelis, Chris, 18, 20
Kramer, Lawrence, 156
Kubernik, Harvey, 75
Kunz, John, 106
Kurtz, Michael, xiii

Lacan, Jacques, 58
lacquer, 42, 81
laptop, 128, 138
latent, 100
lathe, 81, 116
Latour, 117
lawsuits, 25, 34
Leadbelly, 48, 49
Led Zeppelin, 75
Lee, Stewart, 103
Lefebvre, Sam, 104
legacy, recorded, 39, 43n4, 74, 75
legitimacy, 34, 40, 91, 104, 105
leisure, 58, 128
Lennon, John, 30, 34
Levin, Eric, xiii
Levine, George, 7
Lewis, *L'Amore*, 69

Lewis, Furry, 77
Lewisohn, 40, 43n4
liberalism, 24
Liberman, Mark, 72, 73
library, 28, 91, 138, 139
library postmodernization, 138
lifelike, 9, 47
lifestyle, 138, 140
lifeworld, 71, 72
Light in the Attic Records, 69, 136
likeness, 34
limestone, 123
liminal, 73
Linett, Mark, 76
listener, xii, xv, 3, 4, 5, 6, 8, 10, 11, 17, 18, 19, 20, 21n2, 27, 28, 29, 34, 36, 41, 46, 48, 49, 52, 53, 55, 56, 57, 62, 72, 98, 100, 101, 102, 106, 112, 113, 114, 115, 116, 117, 120, 121, 122, 123, 124, 128, 130, 134, 135, 137, 139, 140, 159
listening subject, the, 2
listening subjects, xiv, xv, 5, 19, 20, 41, 112, 113, 114, 116, 119, 120, 123, 124, 127
LITA, 69
literature, xv, 31, 50, 72, 130n5, 142
live music, 5, 15
Lomax, John and Alan, 48
London Records, 12
Long Playing records, 124
Lostpedia, 130n2
loudspeaker, 11, 120, 121
Lovitt, Bryn, xiii
loyalty, 3, 4
LP: 33 1/3 LP, 16; LP records, xii, 12, 13, 16, 19, 25, 27, 28, 30, 41, 43, 47, 49, 53, 55, 57, 69, 71, 77, 78, 84, 98, 124, 130n3, 133, 135, 137
Lukács, 71, 103, 104
Luke, Timothy, 100
The Lumineers, 57
Lunsford, Bascom Lamar, 77
Lyra Atlas, 126
Lyra Atlas, 142
lyrics, 66, 111

machinery of industrial capitalism, 85
Macmillan, Malc, 68
The Matrix, 145

Maerz, Melissa, 57
magnetic tape, 9, 52
magnetic recording, 16
Magnum, Theresa, 1–2
Magoun, Alexander, 3, 6, 9, 16
Maines, Rachel, 128, 129, 130n5
mainstream, xiv, 27, 58, 71, 72, 105, 134, 135; mainstreaming, 49
makers, 13, 142
mankind, 97
manufacturers, 9–12
Mapleshade, 114
Marcus, Greil, 72
Marcuse, Herbert, 72
marginalized, 72
marketable, 49
marketers, 8, 21n2
marketplace, 149
Marsh, Dave, 37
Martin, Andrew, 93, 94
Martin, George, 26, 27, 33, 37, 40
Marx, Karl, 68, 88–89
Marxist, 67
masculinity, 127, 128
mass-cultural, 150
mass-produced, 23, 40, 51, 71, 94, 95
master tapes, 26, 38, 55, 78
mastertape, 43
matrix (recording), 55, 81
The Matrix, 145
matryoshka dolls, 88
McBean, Angus, 23
McCartney, Paul, 37
McDonald, Ian, 27
mechanism, xv, 110, 112, 118, 129
mechanization, 50, 117
mediated, 6, 33, 46. *See also* unmediated
mediation, 6, 10, 13, 19, 33, 41, 115, 146, 147, 150
mediator, 114, 147
megastore, 85
Melody Maker, 124
membership, 150–151
memorabilia, 81, 88
Memphis, 107n3
mercantile, 80
merchandise, 149
metaphysical, 31
Metcalfe, Scott, 14

mica diaphragm, 117
microgroove long player, 9
microphones, 40, 102
microphysics of power, 118
Milano, Brett, 15, 46, 55, 59
Millard, Andre, 124
millennial hipsters, 73
Miller, Scott, 53
Mills, Sara, 11
mimesis, 7, 8, 34; mimetic, 55
minstrel show, 8
minstrelsy, 66–67
miscue, 123
mishearing, 20
Mississippi John Hurt, 77
modern industrial period, 97
modernism, 50, 51
modulations, 9
Moist, Kevin M., 65
Mojo magazine, 69
mold, 35
monaural, 120
Mono, 13, 16, 24, 25, 26, 28, 37, 39, 41, 74, 76, 81, 151; monophonic, 30, 37
mortality, 115, 123
Moses, Morton, 77
mothers (recording), 55
motor, 117
Motown, 25
MP3, 17, 34, 36, 37, 46, 103, 105
MTV *Unplugged*, 48
Muensterberger, Werner, 46
multitrack tapes, 30
multitrack recordings, 38
multivalent, 71, 72, 80; multivalence, 72
Mumford and Sons, 57
museum, 95, 97, 98, 99–100, 134, 136
music boxes, 149
musical performance, 15, 33, 34, 49, 88, 97, 100, 102, 106
musicologist, 66, 136
myth, 8, 11, 96
myth of total cinema, 7, 18, 114

Napier-Bell, Simon, 70
Napster, 17
nerds, 76, 142
The *New York Times*, 25, 26, 27, 29, 88, 98
New Yorker, 92

Index

niche, xi, 28, 97, 157
nickelodeons, 3
Nipper, xiv, 1–7, 20
Nirvana, 48
Nitty Gritty record cleaner, 114, 125, 127
Norm, 109, 120, 122, 124–126, 149, 150, 154; normalizing judgement, 122, 124, 126; normative, 100, 150
Northern Soul, 82
nostalgia, 52, 53, 147; nostalgic, 31
notation (music), 136
Notepad Productions Volume 1, 68
novelty, 9
NPR, 14
NWOBHM, 68
NYU's Clive Davis Institute of Recorded Music, 18

Oatman-Stanford, Hunter, 147
O'Brien, Pat, 13
O'Holla, Sarah, 142–144, 146–148
objectification, 129
objectivity, 10, 11, 64, 74, 103, 104
objet petit a, 58
obsolescence, 95
Oceanic Flight 815, 109
Of Monsters and Men, 57
Old Crow Medicine Show, 57
Olive, Sean, 16, 56
one-to-one correspondence, 18
ontology, 111; ontological, xiv, 36, 51, 84, 145; ontologically, 7, 71. *See also* deontology
opera, 3, 138
Opie, John, 145, 148
oppression, 92; oppressive conditions, 93
Orbitrac pads, 125
orchestra, 9, 29, 121
orchestrating, 31
The Oregonian, xii
originality, 51, 113
Orthophonic, 9
Orville, Miles, 50–51
Osborne, Richard, 7, 16, 124
otherness, 7
outlaw collectors, 74
outmoded, 49, 99
overdubbed, 40
ownership, 36, 37, 71, 101, 102, 107n4

Pandora, 70
panoptic, 5, 129
paradigm, 95
Paramount Records, 87–88, 90, 91
Parks, Van Dyke, 33
Parlophone, 30, 55
parlor, 3, 4
patent, 4, 7
Patke, Rajeev, 36
Paul, Les, 32
Peabody Institute of Johns Hopkins University, 14
pedagogical, 20, 117
Peim, Nick, 36
penny arcades, 3
Peoples, Glenn and Russ Crupnick, xi, xii
Perception, xiv, 6, 8, 14, 16, 18, 49, 56, 130, 156, 160
Perfect Sound Forever, 137
performance, 5–6, 9, 10, 11, 12, 13, 15–16, 30, 32, 33, 34, 35, 36, 40, 41, 42, 47, 48, 49, 51, 55, 57, 59n2, 62, 63, 71, 72, 75, 77, 88, 97, 100, 101, 102, 105, 106, 120, 127
performer, 5, 8, 36, 48, 49, 63, 66, 77, 82, 90, 91, 105, 137, 154, 159
Peters, John Durham, 48
phantom objectivity, 103
phenomenology, 120
Phillips, Marc, 137
Phillips, Sam, 107n3
photography, 7, 10, 11, 51, 128
physical media, 45, 70
piano, 20, 40, 55
pickups, 120
Pink Floyd, 133
piracy, 36, 74
Pitchfork, 88
playback, xv, 3, 5, 10, 21n2, 41, 57, 117, 119, 120
polysemous nature, 4
polysemy, 63
polyvinyl chloride, 73
Pono, 14, 17, 18
"The Poor Audiophile", 128
popular culture, ix, 24, 51, 72, 80, 160
portable, 20; portability, 71
possession, 105; possessing, 4, 47
post-apocalyptic narratives, 56

post-war, 129
post-modern anxieties, 51
power relations, 4, 11, 129
PR push, 27
praxis, 97, 98
prelapsarian, 31, 57; prelapsarian ideal, 31
presence, 6, 24, 28, 31–32, 35, 45, 48, 55, 76, 115
preservation, 62, 67
preservative function, 156
Presley, Elvis, 95, 96
pressing, 13, 23, 36, 41, 42, 43n2, 47, 55, 57, 69, 77, 78–79, 81, 82, 83, 105, 150
pressing plant, xii, 79, 82
prestige, 124, 155
Prial, Dunstan, 93
primitive, 48
Prince Charming, 146
privacy, 149
Pro Tools, 55
producer (record), 21n2, 27, 33, 40, 41, 63, 77, 104, 118, 120
profile (online user), 137, 140, 141, 157
promotional, 79, 137
psych, 80
Public Image Limited, 98
publicity, 4
punishment, 124
Punk Rock, 72, 74; Punk, 68, 71, 72, 74, 135
pushers, 6

race records, 90
racial, 8, 49, 66, 91, 150
racism, 49, 66
Ragtime, 66
Rare Record Price Guide, 81, 82
rarity, 79, 83
rationality, 71, 72
Rawson, Eric, 19, 20, 21n2
RCA Victor, xiv, 93
realism, 5, 6, 7, 9, 17, 18, 50, 51, 112
reality, 62, 96, 100, 102; recorded sound, 6, 7, 17, 18, 50, 130; virtual, 133, 143, 145, 147, 159
rebel, 64, 70
Record Collector, xiv, xv, 23, 43n2, 51, 55, 61, 62, 63, 64, 65, 66, 67, 68, 70, 71, 72, 74, 77, 80, 81, 84, 85, 87, 90, 105, 106, 114, 125
record club, 136
Record Store Day (RSD), xii, xiii, 95, 96, 102, 103, 104, 106
recording engineer, 16, 18, 27, 40, 55, 123
recording studio, 10, 32, 40
recording system, 9
Records My Cat Destroyed, 133
recreation, 7, 36
Red Seal label, 3, 28, 57
Reel Music, 30
reemergence of vinyl, xi, xv, 49, 135, 136, 138
reification, 70, 71, 72, 88, 92, 103, 104
Renaissance of vinyl, xii, 137
resolution, 14, 17, 18, 20, 26
retrogressive, 67
Retromania, 65
Reuters, 37
Revenant Records, 88
reverb, 30
reverberation, 101
Revolution, 56
Revolver, 33, 40
Reynolds, Simon, 65
RHC, 98
RIAA, 67, 71, 74, 80
Richards, Keith, 104
Richardson, Mark, 18, 77
Riesman, 70
Riley, Tim, 32, 33, 40
Ritualize, 115, 137
Robot, 61, 130n1
Rock Band, 159
Rockwell, Ken, 127, 128, 129
Rodgriguez, Sixto, 69
Rolling Stone, 75, 135
Rollins, Henry, 97, 104
Rose, Nikolas, 112
Rosen, Jay, 140
Rosoff, Matt, 45
Rough Trade Records, xiii
Royal Academy, 1
royalty, 69, 91
RPM, 9, 12, 98, 124
RSD; see Record Store Day
run-out groove, 81–82

SACD, 19

sales-tracking, xii
Salvadori, Filippo (4 Men With Beards), 98
San Francisco Weekly, 104
Sax, Doug, 42
scalpers, 104
scarcity (and value), 80, 81
scientific, 10, 14, 100, 120, 128
scratching (DJ), 135
Searle's Chinese Room, 145
Seeger, Pete, 77
self-fashioning, xv, 140, 141, 142
self-presentation, 139, 140, 143
selfie, 149
seller, 82, 83, 85, 93, 149
semiotic, 72, 142
Shah, Neil, xii
Shakur, Tupac, 92
Shankland, Stephen, 18
sheet music publishers, 8
shellac, 1, 29, 46, 66, 67, 71, 93, 109, 112, 116, 123, 124, 149
Sherry, John F., Jr., 4
Shuker, Roy, 62, 63, 74, 77, 81, 137, 138
Silke, John, 15, 16, 134
Silverstone, Roger, 2
simulation, 96, 133
Sinatra, Frank, 25
Slatoff-Burke, Megan, 11
sleevenotes, 87
SME Model 30/12, 129
Smith, Harry, 77–78
Smith, Norman, 40, 42
Smith, Patti, 151
The Smiths, 151
social media, xv, 74, 133, 135, 136, 138, 139, 140, 141, 142, 149, 150, 151, 152, 154, 155, 156
software, 116
soloists, 6
songwriter, 91, 133
songwriting, 37
Sony/ATV Music Publishing, 37
Sony/Phillips, 137
sound performances, 5
sounding 'black', 8
SoundScan, xii, xiii
soundstage, 55, 101–102
Sousa, John Phillips, 6
Sousa marches, 66
souvenir, 63, 135
Souvignier, Todd, 135
speaker, 42, 53, 101, 113, 119, 120, 122, 123, 127, 128, 129, 147
speaker enclosure, 122
specifications, 9
Spector, Phil, 32
Spin magazine, 28, 135
Spizer, Bruce, 78–79, 81
Spotify, 29, 54, 70
Stafford, Jo, 45
Stamper, 52, 55, 81
"Steamboat Willie", 67
Steely Dan, 59n2
Steffen, David, 18
stereophonic records, 9
stereophony, 102
Sterne, Jonathan, xiii, 9, 19, 21n1, 21n3, 25, 33, 34, 36, 37, 48, 49, 101, 102, 106, 112, 113, 114, 117, 119, 123, 130n4, 137, 147, 156; The Audible Past, xiii
Stevehoffman.tv, 47
Stooksbury, Clark, 150, 151
Strangelove, Michael, 156
Straw, Will, 70, 80
Stroh violin, the, 6
Stubbs, David, 24
stylus, 45, 59, 113, 114, 116, 118, 123, 124, 126, 127, 130n6
Suisman, David, 3, 6, 20, 28, 29
Sullivan, Matt, 69
sunshine pop, 80
sweet spot, 114, 120, 129, 130
synthesizer, 57
Sørensen, Knut H., 117

T. Russell, James, 13, 155
talking machine, 2, 3, 5, 7, 8, 9, 16, 28, 118, 123. *See also* Victrola; Victrola XI Talking Machine
Target, 85
taxonomy, 20, 66
Technics, 147
technophilia, 127
teleology, 111
telos, 111
tempo match, 135

terministic screens, 62, 71
textual practices, 146
thingness, 34, 37, 102, 142, 146, 147. *See also* Heidegger
Third Man Records, 87, 88, 90, 92, 94, 95, 96, 136
Thompson, Dave, 94
Thompson, Erik, 52
Thompson, Mark, 140
Threndyle, Steve, xii
tinfoil phonograph, 135
tinny, 16, 28
tonal, 34, 123, 138
Torrent, 70, 79. *See also* digital downloads; download; downloaders; illegal downloading
Touchstone, 130n1
The Tracking Angle, 125
Travis Elborough, xiii, 49, 53
trope, xiv, 8, 48, 143, 145
Trout, 147
Truth and Method, 1
Tumblr, 136
Turkle, Sherry, xv, 141, 144, 145, 149, 150
Turntable: turntablism, 71
Turntable Kitchen, 136
Turtles, 13; *The Turtles! Golden Hits*, 13
Twitter, 79, 136

Universal Music Group (UMG), 28
Uncle Saul, 113
underdetermined, 80
unfamiliar technologies, 2, 46
uniqueness, 32, 35, 71, 145
United Radio and Electrical Workers, 93
unmanly, 127
unmediated sound, 19
unnatural, 122
unreleased recordings, 38
unrestored audio, 95
USB, 128
user profile, 140

Vaher, Berk, 63, 64, 65
validation, 85, 152, 153
valuation, xiv, 85
VariSpeed, 40
vault, 38, 76, 85

VCLT (Vinyl Community Love Train), 155
Vechinski, Matthew James, 107n4, 139, 140, 142
The Velvet Underground, 151
verisimilitude, 19
VHS, 61
vibe, 154, 158n10
Victor Records, 116, 123
The Victor Talking Machine Company, 2, 3, 5, 7, 8
Victorian animal painting, 1
Victorian realism, 7
Victrola, 9, 116–117, 124, 149
Victrola XI Talking Machine, 123
Villchur, Edgar, 120, 122, 123
vintage feel, 114
The Vinyl Countdown, xiii
vinyl enthusiasts, xv, 115
Vinyl Junkies, 46, 58
Vinyl Me, Please, 136
Vinyl Moon, 136
vinyl technology, xi, 147, 148
vinylphile, 47, 55, 126, 128, 129
vinylphilia, 56, 126
virtual reality, 133, 143, 145, 147, 159
virtuosity, 99
vocal, 9, 40, 57
VPI 16.5 record cleaning machine, 114, 127

Waehner, Michael, 15, 47, 124, 125
Wal-Mart, 85
"walker" (*The Walking Dead*), 45, 56
The Walking Dead, 56, 59n1
Walkmans, 134
WALL-E, 61, 85
warmth, 18, 19, 152
Waters, Roger, 133
Watson, Doc, 57
waveforms, 18
wax cylinder, 3, 66, 99
Wayne, Michael, 125–126
webcam, 150
Weinstein, Mark, 136
Weller, Paul, 103
White, Jack, 88, 95, 96
The White Stripes, 88
White Whale, 13

Williams, Alex, 98
Williams, J. Mayo, 90–91
Williams, Raymond, xiv, 56, 58
Willis, Susan, 66–67
Wilson, Brian, 33, 76
Windmüller, Sonja, 95
Winer, Ethan, 14
Wisconsin Chair Company, 90
withering of the aura, 32
Womack, Kenneth, 24, 27
women's voices, 6
Woodbury, 45
World War II, 16, 128, 147

Yochim and Biddinger, 48, 114, 115, 123, 135
Young, Neil, 14, 77
YouTube, 149, 150, 151, 152, 156; YouTube channels, 150; YouTube Vinyl Community, x, 23, 96, 150, 152, 153, 154, 156, 158n2

Žižek, Slavoj, 144–145, 146
Zolten, Jerry, 32, 33
Zwicky, Arnold, 73

About the Author

Paul Winters's love of records goes back to a time in his childhood when his parents gave him a GE Show 'N Tell record player for Christmas. A 45 record with "Yellow Submarine/Eleanor Rigby" on it, gifted to him by a friend, introduced him to the sound of the Beatles and ignited a lifelong passion for vinyl records. As new recording technologies came along, he tried them out, but none of them surpassed the thrill he got from entering a record store or breaking the shrink wrap on a new LP. After receiving a master's degree in English and a PhD in Victorian British literature, he went on to teach a variety of courses, including one called "Technology, Society, and Culture." The research for the course and the experience of teaching it to students interested in technology expanded his scholarly interests to include the intersection between technology and the humanities. It was out of all of these interests combined that Winters began thinking and writing about the reemergence of vinyl records, perhaps seeking to find the source of what fascinated him as a child all those years ago.